THE EINSTEIN DIET

By

Glenn Moran

Featuring Andrea Teja Natures-Rhythm

Einstein's Concentrated Energy: A Universe of Illusion

You are armed with Atomic Power that is unleashed with plants or carbohydrates in a rainbow of colors in a spectrum of illusionary formulas of concentrated energy, as discovered in the early 20th century by Einstein. To illustrate this, the elements of Hydrogen, water, and Nitrogen, a gas that is 73% of the air we breathe, are the atomic structure of the atomic bomb through fission, splitting of the atom. The bomb, when dropped on the Japanese, turned everything into carbon in a wide circumference! In your body, nitrogen, hydrogen, and Oxygen combine to detox your body and exhaust the carbon they create. In essence, it cleans up viruses and other toxins that are making you sick! You either have energy, or you have depleting energy from inactive ingredients, pharmaceuticals, immune-lowering side effects, and from vaccines with noble metals and genetic modification mRNA toxins. Proper nutrition will ignite an immune offense of inherent superpower. In essence, nature is photosynthesized, and Pharmaceuticals are chemically synthesized without illusionary mathematical codes for the integrity and efficiency of the immune system.

"The atomic and subatomic particles concentrate energy from a chemical reaction, as E=mc2, then propel the energy with nuclear power = mc2. The energy a particle has while at rest, is equal to its mass while at rest, times the speed of light (3x108 meters/second) squared. Because the speed of light is a very large number, Einstein's equation says that particles carry a tremendous amount of energy, even while at rest. Thus, any particle with mass possesses a quantity of "rest energy" in addition to the kinetic and potential energies of motion understood as part of classical mechanics. See Einstein's original paper above, or see an English translation, "Does the Inertia of a Body Depend Upon its Energy Content?" You can also hear Einstein in a 1948 talk about his famous equation. Of Einstein's four influential papers that year, mass-energy equivalence most successfully captured the public imagination.

Scientists can use Einstein's equation to determine the energy released from certain chemical and nuclear reactions. In a nuclear fission reaction, the nucleus is busted apart, releasing a tremendous amount of energy, equal to the binding energy that held the nucleus together. It is the release of energy from the nucleus that makes nuclear explosions so much more powerful than conventional explosions. Conventional explosions, such as those involving dynamite, draw their energy from chemical reactions. Molecules—collections of atoms—change form, but the constituent atoms remain intact. While the energy released in fission is tremendous, the energy released in gluing atoms back together, as occurs in fusion reactions, is even greater. While in theory all matter can be converted into energy, in practice it is possible to split only certain isotopes of the heaviest atoms, such as uranium, or the even heavier element plutonium. Likewise, it is only practical to force the lightest of elements to combine. Thus, the fusion weapons designed in the years following the Manhattan Project use isotopes of the lightest element, hydrogen." Manhattan Project: Science > Nuclear Physics > E=MC^2. I want a focused theory that energy is natural from photosynthesized nutrients that, with rest and activity, generates powerful atomic nuclear energy that pills do not because they are not armed with atomic power organic machine of photon illusional mathematical codes.

The document discusses the importance of energy derived from natural sources, particularly plants, and the role of atomic and subatomic particles in biological processes.

Key points include:

Role of Plants: Plants, through photosynthesis, create proteins and other nutrients that are essential for cellular energy and detoxification.

DNA and RNA: RNA encodes proteins, carrying the necessary information for life, and DNA plays a crucial role in translating these genetic instructions into functional proteins.

Detoxification: Detoxing from synthetic chemicals and pharmaceuticals is essential for maintaining energy and health.

Solar Energy: Sunlight and plant-based nutrition are vital for cellular regeneration and energy production.

Intercellular Communication: Proper nutrition, yielding bacteria that are probiotic, and detoxification support efficient intercellular communication and metabolic pathways.

Health Implications: Avoiding processed foods and synthetic chemicals can enhance health and prevent disease by maximizing the body's natural energy processes.

The diet emphasizes the superiority of natural, plant-based nutrition over synthetic alternatives for maintaining health and energy.

"Ordinary matter is composed of atoms, themselves once thought to be indivisible elementary particles. The name atom comes from the Ancient Greek word ἄτομος (atomos), which means indivisible or uncuttable. Despite the theories about atoms that had existed for thousands of years, the factual existence of atoms remained controversial until 1905. In that year, Albert Einstein published his paper on Brownian motion, putting to rest theories that had regarded molecules as mathematical illusions. Einstein subsequently identified matter as ultimately composed of various concentrations of energy." Braibant, Sylvie; Giacomelli, Giorgio; Spurio, Maurizio (2012)." Particles and Fundamental Interactions: An Introduction to Particle Physics (2nd ed.). Springer. pp. 1–3. ISBN 978-94-007-2463-1.

Energy is light being processed for regeneration. Energy uses photons, photosynthesis of plant life, to detoxify our body and turn waste material into carbon waste material! We release carbon dioxide, which is converted back to carbohydrates. The plants are photosynthesis, to give illusionary light codes of mathematical, genetic and phonetic expression! The way you feel and look!

"The Avery–MacLeod–McCarty experiment was an experimental demonstration by Oswald Avery, Colin MacLeod, and Maclyn McCarty that, in 1944, reported that DNA is the substance that causes bacterial transformation, in an era when it had been widely believed that it was proteins that served the function of carrying genetic information (with the very word protein itself coined to indicate a belief that its function was primary). It was the culmination of research in the 1930s and early 20th century at the Rockefeller Institute for Medical Research to purify and characterize the **"transforming principle."** The Avery– MacLeod–McCarty experiment was an experimental demonstration by Oswald Avery, Colin MacLeod, and Maclyn McCarty that, in 1944, reported that DNA is the substance that causes bacterial transformation, in an era when it had been widely believed that it was proteins that served the function of carrying genetic information (with the very word protein itself coined to indicate a belief that its function was primary). It was the culmination of research in the 1930s and early 20th century at the Rockefeller Institute for Medical Research to purify and characterize the "transforming principle."

What this study discovered is that bacteria from nutrition are important to DNA intercellular communication. Your gut bacteria from food sources are dependent on healthy immunity from organic probiotic bacteria. Dieting, then, is essential to the transfer of DNA energy! Fascinating that the protein complex, the phosphate energy highway, carries bacteria to the cells and determines immune strength. Essentially, you are what you consume!

Ribosomal RNA is transcribed from ribosomal DNA (rDNA) and then bound to ribosomal proteins to form small and large ribosome subunits. rRNA is the physical and mechanical factor of the ribosome that forces transfer RNA (tRNA) and messenger RNA (mRNA) to process and translate the latter into proteins.[1] Ribosomal RNA is the predominant form of RNA found in most cells; it makes up about 80% of cellular RNA despite never being translated into proteins itself. Ribosomes are composed of approximately 60% rRNA and 40% ribosomal proteins, though this ratio differs between prokaryotes and eukaryotes. " (1)Davidson, Michael W. (13 November 2015). "Molecular Expressions Cell Biology: Ribosomes." Molecular Expressions. National High Magnetic Field Laboratory.

(2) "Ribosome | Definition, Function, Formation, Role, Importance, & Facts | Britannica." Encyclopedia Britannica. 2024-03-08. Retrieved 2024-03-29.

In the ribosomes of prokaryotes such as bacteria," 16S ribosomal RNA (or 16S rRNA) is the RNA component of the 30S subunit of a prokaryotic ribosome (SSU rRNA). It binds to the Shine-Dalgarno sequence and provides most of the SSU structure.

The genes coding for it are referred to as 16S rRNA genes and are used in reconstructing phylogenies, due to the slow rates of evolution of this region of the gene.[2] Carl Woese and George E. Fox were two of the people who pioneered the use of 16S rRNA in phylogenetics in 1977.[3] Multiple sequences of the 16S rRNA gene can exist within a single bacterium.[4]

(2) Woese CR, Kandler O, Wheelis ML (June 1990). "Towards a natural system of organisms: proposal for the domains Archaea, Bacteria, and Eucarya." Proceedings of the National Academy of Sciences of the United States of America. 87 (12): 4576–4579. Bibcode:1990PNAS...87.4576W. doi:10.1073/pnas.87.12.4576. PMC 54159. PMID 2112744.

(4) Woese CR, Kandler O, Wheelis ML (June 1990). "Towards a natural system of organisms: proposal for the domains Archaea, Bacteria, and Eucarya." Proceedings of the National Academy of Sciences of the United States of America. 87 (12): 4576–4579.

To put it in simple language, bacteria can be toxic as they form on the skin, in the gut, on meat or food exposed to the atmosphere; therefore, packaging and freezing will keep it fresh until you consume it, extracting nutrition that is probiotic or has preservatives, GMO, and metallic solutions that attract toxic bacteria that cause diseases!

This is why anti-bacterial drugs worked until nature overcame the high-powered antibiotic, antibody effect of these pharmaceuticals! Scientists knew the pills would run a course and be ineffective! The problem is the synthetic drug wipes all bacteria, including beneficial bacteria, fueling the DNA process of energizing cells without side effects! As the WHO reports, "A report, Antibacterial agents in clinical development – an analysis of the antibacterial clinical development pipeline, including tuberculosis, launched today by WHO, shows a serious lack of new antibiotics under development to combat the growing threat of antimicrobial resistance." https://www.who.int/news/item/20-09-2017-the-world-is-running-out-of-antibiotics-whoreport-confirms

"Bacteria (/bækˈtɪəriə/ ⓘ; sg.: bacterium) are ubiquitous, mostly free-living organisms often consisting of one biological cell. They constitute a large domain of prokaryotic microorganisms. Typically, a few micrometers in length, bacteria were among the first life forms to appear on Earth, and are present in most of its habitats. Bacteria inhabit the air, soil, water, acidic hot springs, radioactive waste, and the deep biosphere of Earth's crust. Bacteria play a vital role in many stages of the nutrient cycle by recycling nutrients and the fixation of nitrogen from the atmosphere. The nutrient cycle includes the decomposition of dead bodies; bacteria are responsible for the putrefaction stage in this process. In the biological communities surrounding hydrothermal vents and cold seeps, extremophile bacteria provide the nutrients needed to sustain life by converting dissolved compounds, such as hydrogen sulphide and methane, to energy. Bacteria also live in mutualistic, commensal and parasitic relationships with plants and animals. Most bacteria have not been characterized, and there are many species that cannot be grown in the laboratory. The study of bacteria is known as bacteriology, a branch of microbiology. Like all animals, humans carry vast numbers (approximately 10 to13th to 10 to 14th) of bacteria.[4] Most are in the gut, though there are many on the skin" Sender R, Fuchs S, Milo R (19 August 2016). "Revised Estimates for the Number of Human and Bacteria Cells in the Body."

PLOS Biology. 14 (8): e1002533. doi:10.1371/journal.pbio.1002533. ISSN 1545-7885. PMC 4991899. PMID 27541692

The cultivation of this energy is the best and most efficient method, from the consumption of pure energy plants to stimulate intercellular Illusionary Mathematics in the atomic and subatomic electromagnetic concentration of Energy discovered by Einstein. Gut bacteria, as shown, need to be fed nutrients and processed foods, liquor, processed sugars that breed toxic bacteria transfer to DNA that corrupt genetic and phonetic functions, creating disease! In your physical and genetic formations.

Genes in DNA encode protein molecules through a two-step process involving transcription and translation:

"Transcription: The information in DNA is transferred to a messenger RNA (mRNA) molecule. During this process, the DNA of a gene serves as a template for complementary base-pairing, and an enzyme called RNA polymerase II catalyzes the formation of a premRNA molecule. This pre-mRNA is then processed to form mature mRNA, which is a single-stranded copy of the gene.

Translation: The mRNA is then translated into a protein molecule. The mRNA sequence is read by a ribosome, and transfer RNA (tRNA) molecules bring the appropriate amino acids to the ribosome in the correct sequence to form a protein."
https://www.fcav.unesp.br/Home/departamentos/tecnologia/marcostuliooliveira/translati on_-dna-to-mrna-to-protein-_-learn-science-at-scitable.pdf

This process ensures that the genetic information encoded in DNA is accurately expressed as functional proteins, which are essential for carrying out various cellular functions.

From solar radiation visions and protein alkaline, build a complex metabolic fuel that travels at the speed of light. The formula that is developed by plants' life, and bacteria build protein highways and energize the cells' growing body and intellect. The body and soul!

"The genes in DNA encode protein molecules, which are the "workhorses" of the cell, carrying out all the functions necessary for life. For example, enzymes, including those that metabolize nutrients and synthesize new cellular constituents, as well as DNA polymerases and other enzymes that make copies of DNA during cell division, are all proteins.

In the simplest sense, expressing a gene means manufacturing its corresponding protein, and this multilayered process has two major steps. In the first step, the information in DNA is transferred to a messenger RNA (mRNA) molecule by way of a process called transcription. During transcription, the DNA of a gene serves as a template for complementary base-pairing, and an enzyme called RNA polymerase II catalyzes the formation of a pre-mRNA molecule, which is then processed to form mature mRNA (Figure 1). The resulting mRNA is a single-stranded copy of the gene, which must next be translated into a protein molecule." https://www.nature.com/scitable/topicpage/translation-dna-tomrna-to-protein-393/

Each ray of sun and each seed connect this vital pathway with hydrogen, nitrogen and oxygen. These elements combine detoxing poisons from the body to grow with healthy bones and muscles with protein structures, photons of formulas.

The sun has 118 alkaline elements that are the Periodic Table from the fusion of Hydrogen and Helium at the center, and they are number one on the periodic scale. This alkaline formula communicates with trillions of cells in plants and animals. Nature is a very slow formula, and then it can regenerate the side cells, not side effect cells from high-powered plastic technology. All pills list side effects, and there are usually three or more inactive and active ingredients that also affect major organ functions.

"Some of these materials are used to help the medication remain stable and to control absorption when you take the drug. Excipients like gluten, lactose, sugar, and some dyes can cause allergic reactions in some people." https://www.drugs.com/sfx/

Light therapy using Photons from plant photosynthesis of the entire periodic table of helium and hydrogen, the 1 position of the equilibrium of alkaline infusion of the universe is light medicine!

This light medicine is what Einstein discovered in the early 20th century, that he explained concentrated energy is an illusion!

Only plants can project these codes with RNA to DNA to build and regenerate the body and soul of your Light Body!

Here is the chart for essential elements of Light Therapy fused into plants!

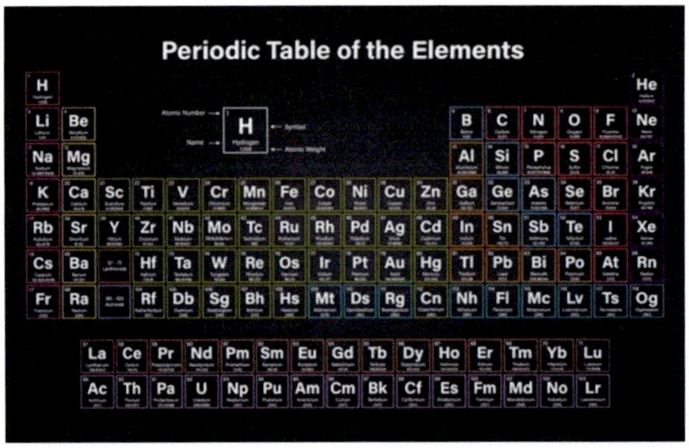

Plastic is a hardened petroleum used in the construction of pharmaceuticals and inhibits oxygen that is a high-power exhaust system, as well as collecting inactively restricting blood flow.

Accumulation of toxins can happen all over the body, leading to disease. Detoxing with plant alkaline is very much like a computer virus. The computer is an algorithm that needs a flow of information. When the information has misinformation, it leaves the pathway of the completed cycle and freezes the computation. "Some of these materials are used to help the medication remain stable and to control absorption when you take the drug. Excipients like gluten, lactose, sugar, and Some dyes can cause allergic reactions in some people." https://www.science.org/doi/10.1126/scitranslmed.aau6753

"Your body reacts, as energy does, with inactive ingredients that cannot compute false information and freeze your energy these toxins grow from a continued poor diet. Older people struggling, as an example, to walk need to flush the inhibiting energy ingredients that are pesticides, chemical synthetic pharmaceuticals, and a plethora of chemical concoctions approved by the FDA. When detoxing and flushing these inactive, sedentary, immune suppressing ingredients and opens metabolic pathways and regenerates energy, releasing power for mobilization of senior citizens, especially. A nutritional diet builds the muscular, skeletal, and organs to efficiently run regenerating the body. Accumulation of inactive sedentary ingredients over time is what people develop into disease. Over 5000 chemicals have been authorized, and this makes detoxing an essential practice for energy levels to increase." https://www.msn.com/en-us/health/other/how-potentially-harmfuladditives-have-infiltrated-america-s-food/ar-AA1uzeKc?ocid=BingNewsSerp

RNA is like ROM in a computer. Plants communicate with RNA that downloads to the DNA helix with organic codes for efficient transfer of biological information. Read Only Memory downloads on the Hard Drive and uses modems to separate the transmissions, giving you windows and channels. They both have virus infections, and both need detoxing to stop freezing up the pathways.

The highly complex DNA helix uses solar radiated photo synthesized food for proper cell transfer in 60 trillion cells. Energy is these photons of nutrition moving from mass to light! To make this light bright is avoiding a poisonous diet that diminishes the Einstein Mathematical Illusion.

The ecological and biological universe is governed by the forces of the elements we breathe and assimilate in light. Within and without are in a connection of light for our survival. Nutrition that is free of poisonous food and fumes, particles, and synthetic chemicals provides a metabolic pathway of energy of illusionary mathematics! This is an exact formula of evolution!

Cultivating this energy, as instructed by renowned spiritual and scientific information, will produce healing through regeneration. Plant medicine uses 118 elements fused by photosynthesis.

Every year, the plants hide in the elements of the earth and await the warmth of spring to bloom and share the solar energy with us and the bees pollinate, making honey to build a protein pathway of energy!

To illustrate this further, you can observe plastic in the encoding of the grocery item. As the computer at checkout scans the bar code, you will, at times, fail to complete the purchase because the plastic has blurred the encoding! The metallic preservative and the plastic-based pharmaceuticals cloud and collect, blocking nutritional metabolic pathways!

Your body has pathways to deliver needed formulas processed by organs to fight disease and is dependent upon energy cultivation. What this energy creates is purely made from unprocessed food that spoils the energy production!

The digestive system of solar radiation photosynthesized plants is a radiation of power to regenerate! Biological integrity!

Another illustration is we inhale Nitrogen, consume Hydrogen and the third to complete energy cycles is Glycol, plant sugars.

When they use fusion, or splitting the atom, Hydrogen and Nitrogen are highly explosive atomic bombs, and Nitroglycerin uses organic sugars to combust. Oxygen aids in the combustion of trillions of energy concentrations.

That is the atomic and sub-atomic power concentrated with illusionary formulas that possess aroma, spectrum colors and chemically produce electromagnetic energy.

1. Electromagnetic forces in the body attract pollination in plant food that constructs a highway of proteins, building an immune defense system!

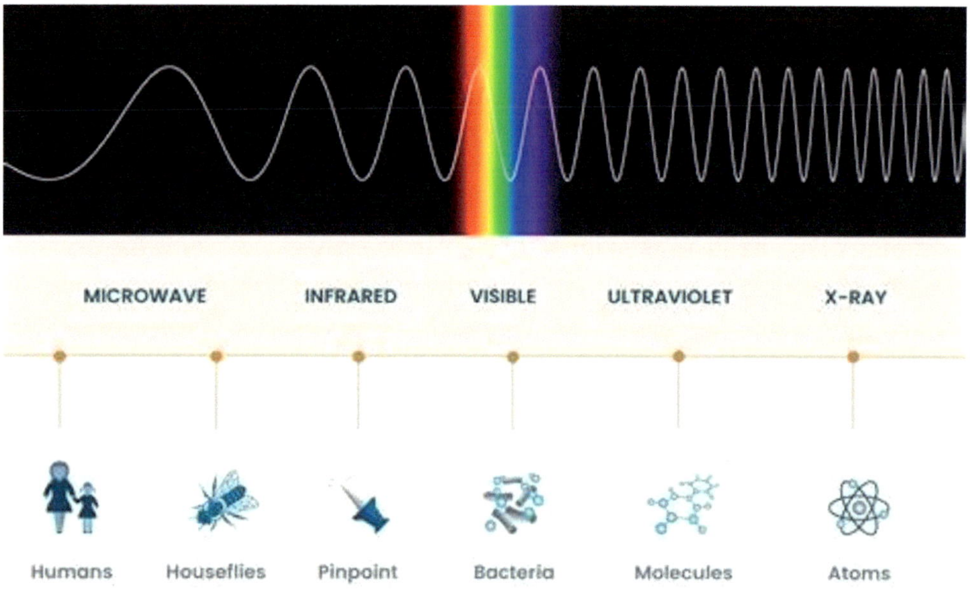

Trillions of energy sparks that travel at the speed of light with energy that sustains life. Intercellular metabolic pathways that have biological, Life-Logic, fuel from solar radiated alkaline photons, organic plants.

This energy was cornered in a market of discovery of pharmaceutical companies bent on profit to discover illusionary formulas but using synthetic test tube or artificial energy. This is why they name numerous side effects from inorganic, inactive ingredients that simply do not compute to organic algorithms. Simply put, Pharma is chemical synthesis, and plants use photosynthesis, one in a test tube, the other by solar energy.

Pollination is the spreading of nutrients essential to a healthy life. THC and CBD are like magnetic attractions to pollination gathering plant photons of millions of plants that is the formula of intercellular communication of protein building functions. It is our personal algorithm and defense system.

The highly sophisticated, intelligent system replicates 60 trillion cells every second in your body. It has the power, when fueled, to eliminate disease by RNA from plants. It's a sugar, Ribose Nucleic Acid, that chemically changes to an acid minus oxidation, Deoxy Ribose Nucleic Acid. It is soaked up on the DNA receptors for instructions on body and intellectual regeneration, moving forward with internal energy.

Pharmaceuticals inhibit energy process lowering immunity with side effects. The biological organic formula is specially created from photosynthesized plant material. Test tube synthetics are designed to overpower the intercellular transmission with a chemical formula using inactive ingredients that are void of solar radiation, which the intercellular DNA does not function biologically, causing the next cell on the side's cells. This is then called side effects! When it is antibiological, not logical to the next cell! It can be an example that if you talk English to a Russian person who does not understand English, the message has the side effect of misunderstanding.

The Helix of the DNA is filled with formula-specific receptors with a nitrogen base building protein. Your body is an amazing complex of codes from nature only!

"This demonstrated that DNA was the material that contained genetic instructions. Specific sequences of nitrogenous bases that code for proteins or regulatory RNA molecules are called genes. Each strand of DNA is like a recipe book for synthesizing proteins. Certain sequences of nitrogenous bases along the strand encode particular RNA molecules. These sequences are called genes. mRNA molecules transcribed from genes are translated into proteins later. A nitrogenous base is an organic molecule that contains nitrogen and has the chemical properties of a base. There are four nitrogenous bases that occur in DNA molecules: cytosine, guanine, adenine, and thymine (abbreviated as C, G, A, and T). RNA molecules contain cytosine, guanine, and adenine, but they have a different nitrogenous base, uracil (U) instead of thymine." https://www.visiblebody.com/learn/biology/dna-chromosomes/dna-structure

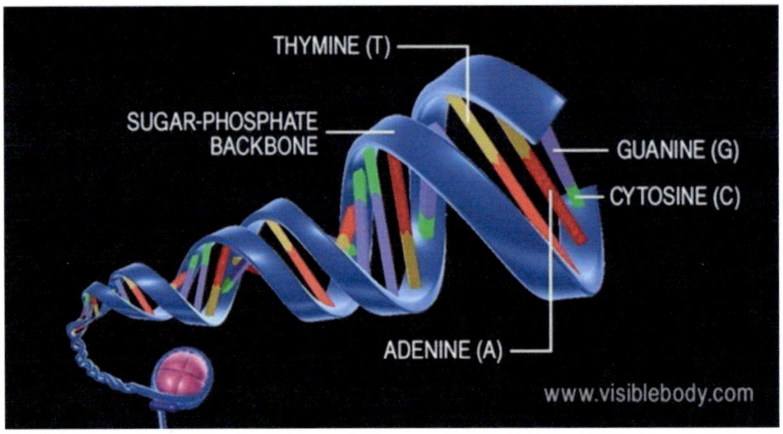

For the body to efficiently run the daily cultivation of Sunlight, Oxygen and Hydration plus nutrients, will detox and build protein structures throughout the body in concentrated energy from subatomic levels, regenerating cells for fresh energizing. This will, over time will maximize healthiness. When there are physical defects, the body will seek a bridge to create pathways. An example is that a blind person has heightened hearing. Autistic scrambling of brain codes will manifest into super intellectual genius in many Autistic people. "While not formally diagnosed, many believe that Albert Einstein had Asperger's. He showed many signs, including some difficulties with small talk." https://hiddentalentsaba.com/geniuseswith-autism/

We all have power unleashed through formulas of RNA and transcribed to DNA. "Molecules of DNA range in length from hundreds of thousands to millions of base pairs. The smallest chromosome in the human genome, Chromosome 21, has around 48 million base pairs." https://www.visiblebody.com/learn/biology/dna-chromosomes/dna-structure

Protein construction maximizes Energic coupled with probiotics and exercise stimulation. The fuel to stimulate this power is best when toxins in mass-produced food are eliminated as much as possible. The FDA is a vehicle to allow mass produced Food and Medicine in separated categories, specifically denying food as medicine! Organic food is solar radiated RNA messaging of illusionary mathematics downloading in highly complex intercellular atomic levels, as well as subatomic energy concentration of energy, as Einstein discovered. **Food is medicine! Food is probiotic fuel! Food is a full spectrum of Solar Energy fused with plants! Solar energy is 118 essential elements for all life!**

This is why a diet enhancing this immense atomic power is a superior medicine and is the only true diet from nature that grew it and is ever present and has endured all civilizations, and I have named it The Einstein Diet!

"Let food be thy medicine and medicine be thy food." Hippocrates

The Einstein Diet

The Einstein Diet is energy maximized from the light energy of plants. Detoxing and toxin avoidance will enhance metabolic brilliance. Feuling the body and mind immune system stimulates intellectual development. To heal, the detoxing will purify the immune formula, which is "The constancy of light" and "The concentration of energy." Einstein

The main concept of the Einstein Diet is to harness photosynthesized nutrients from the light energy of plants. It emphasizes the consumption of organic, plant-based foods that are rich in nutrients and phytochemicals of a highly complex metabolic system. The diet focuses on the principles of biological harmony, physical well-being, and optimal vitality. It promotes that plants are medicine and that exercises combined with a plant-based diet can support detoxification, muscle development, and overall health. The diet also highlights the importance of maintaining a balance between alkaline and acidic levels in the body for optimal functioning.

The diet highlights the importance of consuming a variety of organic plants, including fruits, vegetables, legumes, and whole grains. These plant-based foods provide essential nutrients, such as vitamins, minerals, fiber, and antioxidants, which support overall health and well-being. They also contain phytochemicals, which are bioactive compounds that have been shown to have numerous health benefits, including anti-inflammatory and anticancer properties.

By following the Einstein Diet and incorporating plant-based foods into their diet, individuals can harness solar radiation to support their physical and mental health. The diet recognizes that plants are not only a source of nourishment but also a form of medicine that can contribute to overall well-being.

The formula will produce energy for benefits and give the energy of life for a healthy lifestyle.

- Nutritional Essentials
- A balanced diet is crucial for optimal overall well-being.
- Biological Harmony
- Electromagnetic light waves encode genetics and phonetics to shape the body's form and function.
- Physical Well Being

Aerobic exercise and muscle stretching support detoxification and muscle development.

- Optimal Vitality

Maintaining balance and vitality is reflected in how you feel and appear.
Phyto-Photo evolution is the process by which plants convert light energy into chemical energy through photosynthesis. This energy is then consumed by organisms, including humans, through the consumption of plant-based foods. The energy consumption by humans is directly linked to the energy produced by plants through photosynthesis, and phyto-photo-reevolution plays a crucial role in sustaining energy flow in the ecosystem.

Eating energy with anti-inflammatory Phyto-Photon, Plant-Light, and organic ingredients will chemically synthesize energy-neutralizing toxins. When toxins have no toxic food from toxic ingredients, that is toxic fuel, their bacteria cannot grow, they then are assimilated into alkaline healthy energy and zapped of toxic energy. This is a carbon process of breaking down and regeneration. "The carbon cycle is that part of the biogeochemical cycle by which carbon is exchanged among the biosphere, pedosphere, geosphere, hydrosphere, and atmosphere of Earth. Other major biogeochemical cycles include the nitrogen cycle and the water cycle. Carbon is the main component of biological compounds as well as a major component of many minerals, such as limestone. The carbon cycle comprises a sequence of events that are key to making Earth capable of sustaining life. It describes the movement of carbon as it is recycled and reused throughout the biosphere, as well as long-term processes of carbon sequestration (storage) to and release from carbon sinks."

Libes, Susan M. (2015). Blue planet: The role of the oceans in nutrient cycling, maintain the atmosphere system, and modulating climate change. Archived 8 January 2023 at the Wayback Machine. In: Routledge Handbook of Ocean Resources and Management, Routledge, pages 89–107. ISBN
9781136294822. "Carbon is tetrahedrally bonded to oxygen."

Carbon and oxygen are bonded as the earth regenerates with respiration. By using plants, the carbon cycle and breathing for respiratory function exhaust or exhalation, Volatile Organic Compounds. In other words, the body knows what to do with toxins, and the carbon process is exhaled for recycling. (The Pranayama chapter in this book will use this process of respiratory exercising.)

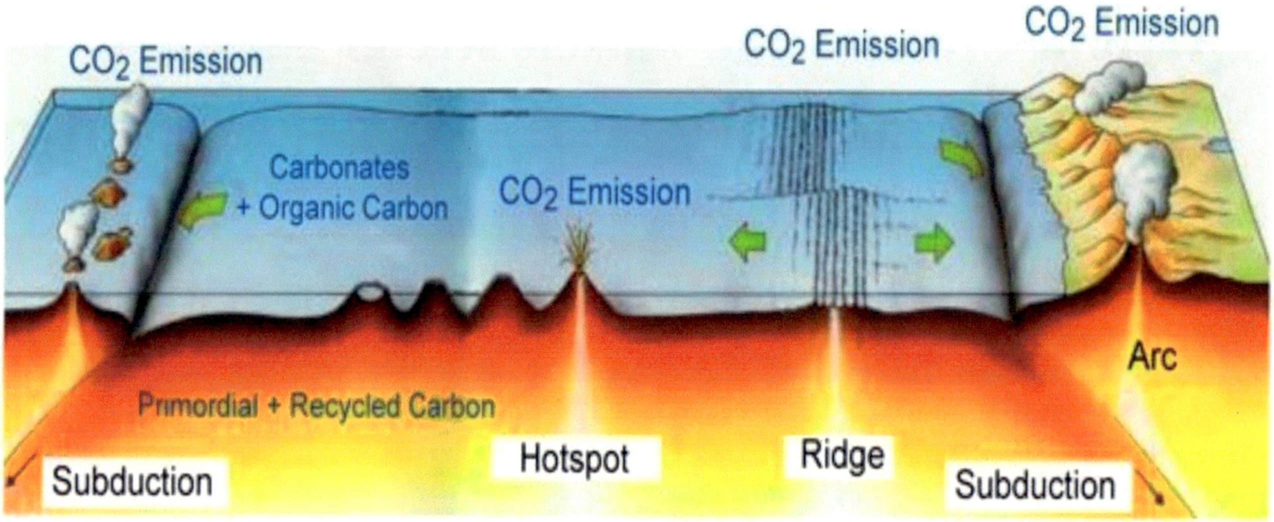

Carbon cycle. (2024, March 12). In Wikipedia. https://en.wikipedia.org/wiki/Carbon_cycle

The Einstein diet follows the path of breakdown to carbons with cycles of four main sources. Hydrogen, Oxygen, Nitrogen and Carbons. When you have organic levels of fuel, the formula combines to fertilize and detox the metabolic system. When you eat inactive ingredients, it becomes a variable diminishing energy production. Inactive ingredients do not break down. The mass-produced food depends on stabilization of the product on inactive ingredients, clogging over time the lumens or artery and vein networks. Organic food is free of debilitating metals, synthetic phosphates, synthetic enzymes and dyes that enhance flavor, color, form and taste.

Organic Phosphates aid metabolic activity powerfully in triphosphate then reduce the speed at which they enter cells. They carry nutrients with them. Nutrients produce energy, protein complexes build immune forces and constructions encoding genetic and phonetic information, how you look, and expression. It is a blueprint with trillions of cells and 500 billion red blood cells feeding them every second, hour, and life.

"The body is a complex organism, and as such, it takes energy to maintain proper functioning. Adenosine triphosphate (ATP) is the source of energy for use and storage at the cellular level. The structure of ATP is a nucleoside triphosphate, consisting of a nitrogenous base (adenine), a ribose sugar, and three serially bonded phosphate groups. ATP is commonly referred to as the **"energy currency"** of the cell, as it provides readily releasable energy in the bond between the second and third phosphate groups. In addition to providing energy, the breakdown of ATP through hydrolysis serves a broad range of cell functions, including signaling and DNA/RNA synthesis. ATP synthesis utilizes energy obtained from multiple catabolic mechanisms, including **cellular respiration**, beta-oxidation, and ketosis." https://www.ncbi.nlm.nih.gov/books/NBK553175/

Vitals

"Vitamins, any of several organic substances that are necessary in small quantities for normal health and growth in higher forms of animal life. Vitamins are distinct in several ways from other biologically important compounds such as proteins, carbohydrates, and lipids. Although these latter substances also are indispensable for proper bodily functions, almost all of them can be synthesized by animals in adequate quantities." https://www.britannica.com/science/vitamin The vitamins

"Water-soluble

Thiamin, Vitamin B1 component of a coenzyme in carbohydrate metabolism; supports normal nerve function, impairment of the nerves, and heart muscle wasting

Riboflavin Vitamin B2 component of coenzymes required for energy production and lipid, vitamin, mineral, and drug metabolism; antioxidant inflammation of the skin, tongue, and lips; ocular disturbances; nervous symptoms niacin nicotinic acid, nicotinamide component of coenzymes used broadly in cellular metabolism, oxidation of fuel molecules, and fatty acid and steroid synthesis skin lesions, gastrointestinal disturbances, nervous symptoms

Vitamin B6, pyridoxine, pyridoxal, pyridoxamine component of coenzymes in the metabolism of amino acids and other nitrogen-containing compounds; synthesis of hemoglobin, neurotransmitters; regulation of blood glucose levels, dermatitis, mental depression, confusion, convulsions.
anemia folic acid, folate, folacin, pteroylglutamic acid, component of coenzymes in DNA synthesis, metabolism of amino acids; required for cell division, maturation of red blood cells, impaired formation of red blood cells, weakness, irritability, headache, palpitations, inflammation of the mouth, neural tube defects in fetus

Vitamin B12, cobalamin, cyanocobalamin cofactor for enzymes in the metabolism of amino acids (including folic acid) and fatty acids; required for new cell synthesis, normal blood formation, and neurological function, smoothness of the tongue, gastrointestinal disturbances, nervous symptoms

pantothenic acid as a component of coenzyme A, essential for metabolism of carbohydrate, protein, and fat; cofactor for elongation of fatty acids, weakness, gastrointestinal disturbances, nervous symptoms, fatigue, sleep disturbances, restlessness, nausea

biotin cofactor in carbohydrate, fatty acid, and amino acid
metabolism, dermatitis, hair loss, conjunctivitis, neurological symptoms

Vitamin C, ascorbic acid antioxidant; synthesis of collagen, carnitine, amino acids, and hormones; immune function enhances absorption of non-heme iron (from plant foods), swollen and bleeding gums, soreness and stiffness of the joints and lower extremities, bleeding under the skin and in deep tissues, slow wound healing, animea.

Fat-soluble

Vitamin A retinol, retinal, retinoic acid, beta-carotene (plant version), normal vision, integrity of epithelial cells (mucous membranes and skin), reproduction, embryonic development, growth, immune response, ocular disturbances leading to blindness, growth retardation, dry skin, diarrhea, vulnerability to infection.

Vitamin D calciferol, calatriol (1,25-dihydroxy vitamin D1 or vitamin D hormone), cholecalciferol (D3; plant version), ergocalciferol (D2; animal version), maintenance of blood calcium and phosphorus levels, proper mineralization of bones, defective bone growth in children, soft bones in adults.

Vitamin E alpha-tocopherol, tocopherol, tocotrienol antioxidant; interruption of free radical chain reactions; protection of polyunsaturated fatty acids, cell membranes, peripheral neuropathy, and breakdown of red blood cells.

Vitamin K phylloquinone, menaquinone, menadione, naphthoquinone, synthesis of proteins involved in blood coagulation and bone metabolism, impaired clotting of the blood and internal bleeding."
https://www.britannica.com/science/vitamin

Several specific micronutrients are required for energy metabolism and ATP production. These include:

B vitamins: B vitamins, such as B1 (thiamine), B2 (riboflavin), B3
(niacin), B5 (pantothenic acid), B6 (pyridoxine), B7 (biotin), B9 (folate), and B12 (cobalamin) play essential roles in energy production. They are involved in converting food into ATP and are necessary for the metabolism of carbohydrates, fats, and proteins.

Vitamin C: Vitamin C is an antioxidant that helps protect cells from oxidative stress. It also plays a role in the synthesis of carnitine, a compound involved in energy production.

Vitamin E: Vitamin E is another antioxidant that helps protect cells from oxidative damage. It is involved in the transport of electrons in the electron transport chain, a process that generates ATP.

Selenium: Selenium is a trace mineral that acts as a cofactor for enzymes involved in energy metabolism. It is necessary for the proper functioning of the thyroid gland, which regulates metabolism.

Zinc: Zinc is involved in numerous metabolic processes, including energy production. It is a cofactor for enzymes involved in carbohydrate and protein metabolism.

Coenzyme Q10: Coenzyme Q10 is a compound that plays a crucial role in the electron transport chain, where it helps generate ATP. It also acts as an antioxidant.

Lipoic acid: Lipoic acid is a compound that is involved in energy metabolism. It acts as a cofactor for enzymes that convert carbohydrates into ATP.

Several specific micronutrients are required for energy metabolism and ATP production. These include:

B vitamins: B vitamins, such as B1 (thiamine), B2 (riboflavin), B3 (niacin), B5 (pantothenic acid), B6 (pyridoxine), B7 (biotin), B9 (folate), and B12 (cobalamin) play essential roles in energy production. They are involved in converting food into ATP and are necessary for the metabolism of carbohydrates, fats, and proteins.

Vitamin C: Vitamin C is an antioxidant that helps protect cells from oxidative stress. It also plays a role in the synthesis of carnitine, a compound involved in energy production.

Vitamin E: Vitamin E is another antioxidant that helps protect cells from oxidative damage. It is involved in the transport of electrons in the electron transport chain, a process that generates ATP.

Selenium: Selenium is a trace mineral that acts as a cofactor for enzymes involved in energy metabolism. It is necessary for the proper functioning of the thyroid gland, which regulates metabolism.

Zinc: Zinc is involved in numerous metabolic processes, including energy production. It is a cofactor for enzymes involved in carbohydrate and protein metabolism.

Coenzyme Q10: Coenzyme Q10 is a compound that plays a crucial role in the electron transport chain, where it helps generate ATP. It also acts as an antioxidant.

Lipoic acid: Lipoic acid is a compound that is involved in energy metabolism. It acts as a cofactor for enzymes that convert carbohydrates into ATP.

Carnitine: Carnitine is an amino acid derivative that plays a role in the transport of fatty acids into the mitochondria, where they are used for energy production.

Nitrate: Nitrate is a compound found in certain vegetables, such as celery, cress, and spinach. It is involved in the production of nitric oxide, which plays a role in energy metabolism and blood flow regulation.

These micronutrients are essential for the proper functioning of metabolic pathways and the production of ATP, the main source of energy in cells.

In this mega speed, the intercellular function is the Key to the immunity function called ATP. "The binding of a divalent cation, almost always **magnesium**, strongly affects the interaction of ATP with various proteins." Storer, A.; Cornish-Bowden, A. (1976). "Concentration of MgATP2− and other ions in solution. Calculation of the true concentrations of species present in mixtures of associating ions." Biochem. J. 159 (1): 1–5. doi:10.1042/bj1590001. PMC 1164030. PMID 11772.

"Nuts and seeds may be small in size, but they pack a nutritional punch with gobs of protein, fiber, healthy fats, and minerals such as magnesium, says Taylor. Here are five options with their **stellar magnesium resumes**:

- "Almonds (roasted): 1 ounce (oz) = 80 mg of magnesium.
- Cashews (roasted): 1 oz = 72 mg of magnesium.
- Flaxseed (whole): 1 tablespoon = 40 mg of magnesium.
- Peanuts (dry roasted): 1 oz = 49 mg of magnesium.
- Pumpkin seeds (hulled, roasted): 1 oz = 150 mg of magnesium.

- Chia seeds: 1 oz = 111 mg of magnesium." https://health.clevelandclinic.org/foods-that-are-high-inmagnesium

When the ATP is bound to proteins, a complex of energy begins when the Triphosphate forms, it reaches the E=MC squared speed that converts the mass of nutrients into Energy." Organic phosphates are commonly found in the form of esters as nucleotides (e.g., AMP, ADP, and ATP) and in DNA and RNA." "These phosphorylation and dephosphorylation reactions are the immediate storage and source of energy for many metabolic processes. ATP and ADP are often referred to as high-energy phosphates, as are the phosphagens in muscle tissue. Similar reactions exist for the other nucleoside diphosphates and triphosphates." https://en.wikipedia.org/wiki/Phosphate

The energy then changes from a Triphosphate (ATP) to a Diphosphate and into an Uno phosphate that returns to the mass of nutrients, and the Protein Pumps function of the cell repeats the process. This science is omitted by official documents because the AMA says it moves too fast for measurement, therefore negating the use of organic proteins as a healing source of energy protein complexes. The liver is the production point of this energy production, and some pills shut down this and replace it with synthetic, powerful chemicals that corrode the energy production and are referred to as side effects. "Hyperphosphatemia, or a high blood level of phosphates, is associated with elevated mortality in the general population. The most common cause of hyperphosphatemia in people, dogs, and cats is kidney failure. In cases of hyperphosphatemia, limiting consumption of phosphate-rich foods, such as some meats and dairy items and foods with a high phosphate-to-protein ratio, such as soft drinks, fast food, processed foods, condiments, and other products containing phosphate-salt additives, is advised." Renal Dietitian Team, reducing phosphate in your diet, Oxford University Hospitals NHS Foundation Trust, 2022 review.

"ATP = ENERGY. ATP is the main source of energy for the body. Every cell uses ATP. ATP captures chemical energy from the breakdown of food molecules and releases it to fuel other cellular processes.

For maximum ATP production, it is essential to support and optimize mitochondrial function. For example, stress, certain medications, pesticides, and imbalanced gut bacteria composition can all impair ATP production.

For optimal function, mitochondria require vitamins, minerals, enzymes, cofactors, **polyphenols,** and other nutrients (Kramer P, 2017). Adequate nutrition is essential for ATP production and mitochondrial function."
https://mindbodyfunctionalmedicine.com/blog/sports-nutritionhttps://mindbodyfunctionalmedicine.com/blog/sports-nutrition-atp-creation-how-to-create-cellular-energy-for-sports/atp-creation-how-to-create-cellular-energy-for-sports/

Several specific micronutrients are required for energy metabolism and ATP production (Wesselink E, 2019). These include B vitamins, vitamin C, E, selenium, zinc, coenzyme Q10, melatonin, carnitine, nitrate, lipoic acid and taurine for mitochondrial function (Wesselink E, 2019). Vitamin D is also important. In people with low levels of serum vitamin D, vitamin D supplements make mitochondria work better (Kramer P, 2017).
Vitamin B3 helps to stimulate mitochondrial function (Kramer P, 2017). Deficiencies in the micronutrients involved in mitochondrial metabolism are common in our modern diet." (Wesselink E, 2019).

"Types of **Polyphenols**:

Flavonoids: Found in fruits, they include compounds like quercetin and catechins.

Phenolic acids: Present in vegetables, whole grains, and fruits like apples, blueberries, and cherries.

Polyphenolic amides: Found in chili peppers.

Others: Resveratrol in red wine and ellagic acid in berries.

Health Benefits of Polyphenols:

Heart Health: Polyphenols can help manage blood pressure and maintain healthy blood vessels, reducing the risk of heart disease.

Diabetes Risk Reduction: They help control blood sugar levels and improve insulin sensitivity.

Anticancer Properties: Some polyphenols may block tumor growth and kill cancer cells.

Immune System Support: Polyphenols activate the immune system and promote a healthy gut microbiome.

Foods Rich in Polyphenols:

Berries: Blueberries, strawberries, and blackberries.

Apples: A great source of flavonoids.

Dark Chocolate: Cocoa is rich in polyphenols.

Tea: Especially black tea.

Coffee: Contains phenolic acids.

Vegetables: Broccoli, carrots, and chili peppers.

Whole Grains: Lignans and stilbenes found in grains.

Spices: Cumin and cinnamon."
https://www.bing.com/search?q=polyphenols+foods&pc=S125&form=S1GDAG&ptag=N5103&showconv=1

All living things use ATP. We need it to fuel all activities and processes of the human body, including, of course, participating in sports."

The sages meditated for thousands of years, and the plants and the exercises enlightened them, then Chi and Yoga were created, especially the medicine from plants. The meditation revealed the inner self, and the ego was dissolved, so your soul united with God.
The energy and spiritual growth were powerful enough to heal. When Jesus was asked about healing and especially raising people from death, he said you must "Fast and Pray a whole lot for that." Bible.

Now science has developed to support the cultivated energy in your body. It was in the 19th Century that immunity was scientifically proven. Elie Metchnikoff was the Father of Immunity. He isolated diseased cells with yeast and was able to prove the alkaline effect of disease neutralization. Alkaline is a plant kingdom, and it is needed to control Acid in a PH level of equilibrium. "Metchnikoff was the first to discover a process of immunity called phagocytosis and the cell responsible for it, called a phagocyte, specifically a macrophage, in 1882. This discovery turned out to be the major defense mechanism in innate immunity, as well as the foundation of the concept of **cell-mediated immunity**, while Ehrlich established the concept of humoral immunity to complete the principles of the immune system. Their works are regarded as the foundation of the science of immunology." Kaufmann, Stefan H E 369807

Elie Metchnikoff proved that macrophages play a key role in the immune response by demonstrating their ability to engulf and destroy foreign particles, such as bacteria, through phagocytosis. His work laid the foundation for our understanding of innate immunity and the importance of macrophages in defending against infections." Immunology's foundation: the 100-year anniversary of the Nobel Prize to Paul Ehrlich and Elie Metchnikoff."

Nature Immunology.

"His theory, that certain white blood cells could engulf and destroy harmful bodies such as bacteria, met with skepticism from leading specialists, including Louis Pasteur, Emil von Behring, and others. At the time, most bacteriologists believed that white blood cells ingested pathogens and then spread them further through the body. His major supporter was Rudolf Virchow, who published his research in his Archive für pathologische Anatomie und Physiologie und für klinische Medicin (now called the Virchows Archiv).[27] His discovery of these phagocytes ultimately won him the Nobel Prize in 1908." "The Nobel Prize in Physiology or Medicine 1908." Nobelprize.org. Nobel Media AB. Retrieved 16 March 2015.

This shows that Fast Food or processed foods and
Pharmaceuticals make you manifest disease and addict you to hyper levels of chemicals, producing additional diseases that are called comorbidities. The CDC said the COVID Vaccine killed people with 5 or more comorbidities. It was the shot that killed them as the mRNA invaded the T-cell immunity function and attacked the spine and blood groups of the bones. This is based on blood-clotting deaths that were happening right after shots everywhere but are reported as COVID mortality from COVID-19.

"What You Need to Know

A person with any of the medical conditions listed below is more likely to get very sick with COVID-19. If you have one of these conditions, talk with your healthcare provider about how best to protect yourself from severe illness from COVID-19." CDC

The main levels of medicine had lines drawn between the two studies. One was holistic and used plant-based medicine, Marijuana, being a main use for its high level of anti-inflammatory healing power. The other was Alchemy, which was mainly apparatus gadgets and chemical inventions geared for monetary levels, including the reproduction of gold.

Opium was a painkiller in many areas of the world, yet was not used much until WW1, which included morphine, and the addiction rate was high. This highly addictive overuse and synthetic lab-created medicine replaced organic medicine, and heroin has killed millions now from Pharmaceuticals. The AMA and FDA were appointed by Rockefeller. They then outlawed any holistic healing that had been proven in the late 19 century. What Elie Metchnikoff proved is that macrophages, or Big Eaters, neutralize diseases. Only a holistic diet can be energy-efficient from the light in plants for regeneration by neutralizing toxic growth in cells.

This diet is holistic, and the formula of Einstein's is Energy = Mass X the Speed of Light Squared. "Plants have photons from solar radiation. A photon is an elementary particle that is a quantum of the electromagnetic field, including electromagnetic radiation such as light and radio waves, and the force carrier for the electromagnetic force. "**Photons are massless, [a] so they always move at the speed of light in a vacuum,"** 299792458 m/s (or about 186,282 mi/s). Joos, George (1951). Theoretical Physics. London and Glasgow: Blackie and Son Limited. p. 679.

"The modern photon concept originated during the first two decades of the 20th century with the work of Albert Einstein, who built upon the research of Max Planck. While trying to explain how matter and electromagnetic radiation could be in thermal equilibrium with one another, Planck proposed that the energy stored within a material object should be regarded as composed of an integer number of discrete, equal-sized parts. To explain the photoelectric effect, Einstein introduced the idea that light itself is made of discrete units of energy." "Gilbert N. Lewis." Atomic Heritage Foundation. Retrieved 2019-03-09.

To fuel this energy, a Mega Speed has been evolving from fire into water. From Fossil Fuel to Protein Electromagnetic. This world is filled with spectra and frequencies that send the earth's minerals and then photosynthetic function for plants. This fuels the metabolic path to your DNA from probiotic foods or organic plants, producing RNA for instructions for a healthy metabolic system.

This Mega Speed loads 50,000,000 codes in a rapid sequence, energy pumping trillions of cells. The liver produces the Key to the trillions of cells with a phosphate called ATP. The ATP Phosphate is energy in making as the mass of nutrients is converted to energy that is transcribed to DNA from RNA messengers.

The trillion of cells every second are the massive energy pathway of your body. This is the Einstein Diet from Phyto-Photo energy consumption from mass to energy. Here is the electromagnetic and particles radiated in a geometric equation.

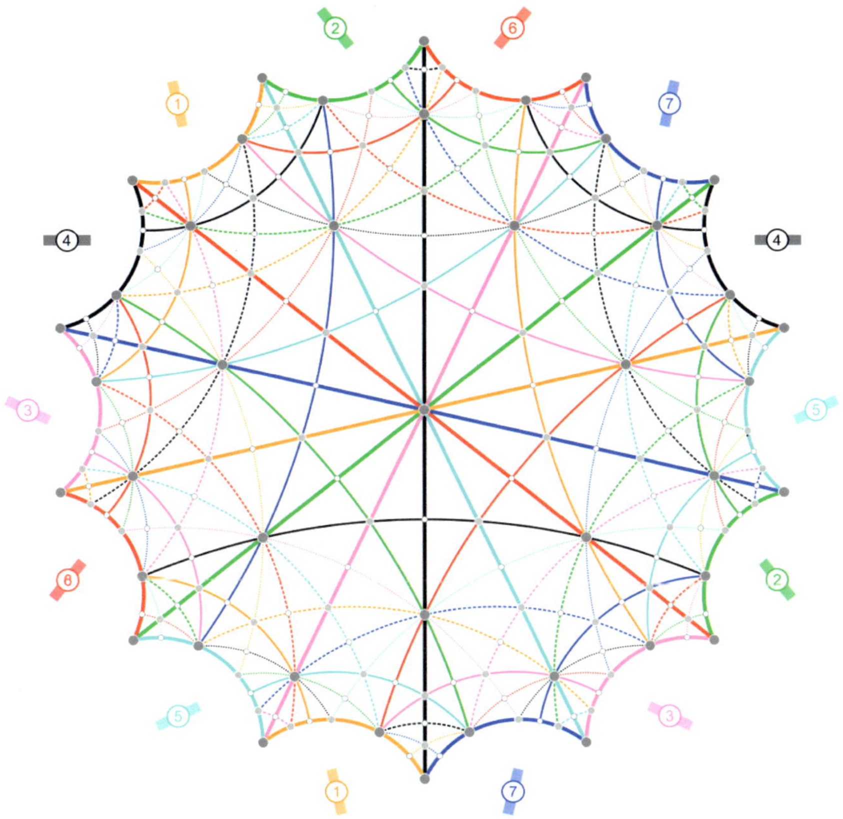

This is a geodesic equation that was Einstein's general theory of relativity. The Einstein field equations (EFE; also known as Einstein's equations) relate the geometry of spacetime to the distribution of matter within it. "The solutions of the EFE are the components of the metric tensor. The inertial trajectories of particles and radiation (geodesics) in the resulting geometry are then calculated using the geodesic equation." Einstein, Albert (1916). "The Foundation of the General Theory of Relativity." Annalen der Physik. 354 (7): 769. Bibcode:1916AnP...354..769E. doi:10.1002/andp.19163540702. Archived from the original (PDF) on 2012-02-06.

Some inflammatory diseases associated with vaccine administration include vaccine-induced immune thrombotic thrombocytopenia (VITT), Guillain-Barré Syndrome (GBS), hypersensitivity reactions, acute disseminated encephalomyelitis (ADEM), neuromyelitis optica spectrum disorders (NMOSD), and transverse myelitis." Cureus. 2023 Mar; 15(3): e36323.Published online 2023 Mar 18. doi: 10.7759/cureus.36323

Notice science talks about the spectrum of color and frequencies! **"neuromyelitis optica spectrum"** Nerves with optimum spectrum of colors and frequencies.

When the DNA is from sediment or Petro-energy, the dissolution is not from the sun, water, and oxygen for the purity of RNA. This is an inactive energy source, biologically inept for your body, causing side effects.

To understand that RNA is instructions for DNA to build immunity and your genetic and your phonetic expression. Variables evolved science of biological instructions from photosynthesis of

Solar Radiation and synthesized minerals from earth, and cultivated the plants, and your body is a health regenerative or renewable energy.

Processed Food is constructed with hyper levels of gluten, phosphates, processed sugars, and preservatives. Pharmaceuticals are hyper levels of immune system codes. This then is not an equilibrium of balance of nutrients that fine-tune your nervous system, immune system, and construction with protein metabolic exactness.

This is documented many times as the multiple side effects and corruption of the gastrointestinal destroying immune process through corruptive sources of inactive ingredients.

You need balanced Alkaline to Acid levels or PH levels. The sun has
118 elements that are dedicated to the spread of alkaline life on Earth.

This Phyto-Photo evolution of energy depends upon the consumption of organic matter, food, and the full compass movements of Yoga, full range stretching, exercising plus active lifestyle to move with energy cycles.

Since plants are anti-inflammatory and disease is inflammatory, eating plants and a full spectrum of colors that have been solar radiated will maximize light energy anti-inflammatory regeneration.

Buy organic berries and greens. Eat as much organic food as you can find. Read for inactive ingredients, preservatives, and phosphates that are at hyper levels from synthetic chemical construction. Many times, they are in Latin to hide the translation. Organic food will use proteins that are metabolically efficient of radiated light from the sun and minerals converted to vitamins of the earth. Remember, we are children of the plants that are children of the earth and are children of the stars. "The young Earth cooled enough in tens of millions of years for liquid water to condense, forming the warm, shallow ponds that Darwin first envisioned as life's cradle. Even then, a constant barrage of asteroids pounded Earth's surface, vaporizing the oceans, and sterilizing the planets." https://sciworthy.com/whyhttps://sciworthy.com/why-did-a-bunch-of-asteroids-hit-the-earth-billions-of-years-ago/did-a-bunch-of-asteroids-hit-theearth-billions-of-years-ago/#:~:text=The%20young%20Earth%20cooled%20enough,oceans%20and%20sterilizing%20the%20planet

This light is also produced in your body to convert food to energy, or the mass of food to energy. We now embark upon uncharted waters because pharmaceutical companies cannot stop the universal energy that is superior to any level of medicine. The Cultivation of God's Energy, featuring the physics of Albert Einstein. Einstein won the Nobel Prize, which proved light permeates all matter and reflects light. So, Holistic medicine travels at the speed of $E=mc^2$, and all matter is converted to energy.

"You'll find an excellent example of this in any driver's education manual: If you double your speed, the braking distance is four times longer, so the braking distance is equal to the speed squared [source: UNSW Physics: Einstein light].

The speed of light squared is a colossal number, illustrating just how much energy there is in even tiny amounts of matter.

"A common example of this is that 1 gram of water — if its whole mass were converted into pure energy via $E=mc^2$ — contains energy equivalent to 20,000 tons (18,143 metric tons) of TNT exploding. That's why such a small amount of uranium or plutonium can produce such a massive atomic explosion.

Einstein's equation opened the door for numerous technological advances, from nuclear power and nuclear medicine to the inner workings of the sun. It shows us that matter and energy are one." https://science.howstuffworks.com/science-vsmyth/everydaymyths/einstein-formula.htm

To create this energy at a pristine level, detoxing the body is a must. This is done with Carbohydrates or hydrated carbons. All matter interchanges from energy to mass and then to energy, as discussed. Variables like pesticides and preservatives are released as hydrocarbons as matter breaks down, the regeneration of the planet fuses water, solar radiation, and oxygen, resurrection of plants or Hydroborates, fruits, and vegetables.

The proper diet is based on a full spectrum of colors, and some of the plants have high immune-strengthening properties, but all plants have one thing in common. They are anti-inflammatory, and all diseases are inflammatory.

When the fire created sediment, the source of fossil fuel, oxygen, solar radiation, and water dissolved matter and extinguished the fire, resurrecting as protein energy that is renewable.

Pharmaceuticals use fossil fuels for energy, and Holistic uses protein energy. Proteins from plant pathways are fresh proteins from digestion, and meat is fermented proteins that are higher in acid for digestion. So, a nutritious diet filled with immense speed and power can energize with lasting, renewable strength.

How powerful are nutrients? "The speed of light squared is a colossal number, illustrating just how much energy there is in even tiny amounts of matter.

Put the power of creation flowing through your body!

- Packed with antioxidants, boasting three times the antioxidants of other well-loved berries.
- A natural aid for exercise recovery, helping your body repair and rejuvenate faster.

Dietary Chart

Organic Turmeric

"Turmeric, a spice that has long been recognized for its medicinal properties, has received interest from both the medical/scientific world and from culinary enthusiasts, as it is the major source of the polyphenol curcumin. It aids in the management of oxidative and inflammatory conditions, metabolic syndrome, arthritis, anxiety, and hyperlipidemia. It may also help in the management of exercise-induced inflammation and muscle soreness, thus enhancing recovery and performance in active people. In addition, a relatively low dose of the complex can provide health benefits for people who do not have diagnosed health conditions. Most of these benefits can be attributed to its antioxidant and anti-inflammatory effects. Ingesting curcumin by itself does not lead to the associated health benefits due to its poor bioavailability, which appears to be primarily due to poor absorption, rapid metabolism, and rapid elimination. There are several components that can increase bioavailability. For example, piperine is the major active component of black pepper and, when combined in a complex."

Piperine, "Piperine is the primary ingredient that distinguishes black pepper and other related peppers (like white pepper). The chemical makeup of piperine is C17H19NO3, or 17 parts carbon, 19 parts hydrogen, one part nitrogen, and three parts oxygen." https://sciencing.com/chemicals-salt-pepper-sugar-8506178.html

"Curcumin has received worldwide attention for its multiple health benefits, which appear to act primarily through its antioxidant and anti-inflammatory mechanisms. These benefits are best achieved when curcumin is combined with agents such as piperine, which increase its bioavailability significantly.
Research suggests that curcumin can help in the management of oxidative and inflammatory conditions, metabolic syndrome, arthritis, anxiety, and hyperlipidemia. It may also help in the management of exercise-induced inflammation and muscle soreness, thus enhancing recovery and subsequent performance in active people. In addition, a relatively low dose can provide health benefits for people who do not have diagnosed health conditions."
https://www.ncbi.nlm.nih.gov/pmc/articles/PMC5664031/#:~:text=Research%20suggests%20that%20curcumin%20can,subsequent%20performance%20in%20active%20people

"Curcumin has been shown to increase bioavailability by 2000%. Curcumin, combined with enhancing agents, provides multiple health benefits. The purpose of this review is to provide a brief overview of the plethora of research regarding the health benefits of curcumin."
https://www.ncbi.nlm.nih.gov/pmc/articles/PMC5664031/

Organic plants are free of toxic additives that are inactive energy, or inactive ingredients, depleting immunity energy productions, which are the purest form of maximizing energy-efficient metabolism.

- **Turmeric and Chili Peppers** are powerful anti-inflammatories.
- Terpenes in CBD and THC plants are internationally also known for their medicinal high quality.

Organic Beans

Beans are a strong, plant-based source of protein, fiber, iron, and vitamins that offer many health benefits.

- lima beans
- black beans
- black-eyed peas
- soybeans
- kidney beans

- garbanzo beans
- navy beans
- pinto beans
- red beans

I like pinto beans. The Mexican cuisine is freshly cooked. They use very healthy spices, and because they grow with much sunlight, they are tough on inflammation. Notice when you eat hot peppers, your sinus excrete so much that you need a hanky, right? 3.5 grams of Jalapeño has 150% of the daily requirements of vitamin C. It is filled with photons that follow pain pathways reacting to "Capsaicin, the main pungent ingredient in 'hot' chili peppers, elicits a sensation of burning pain by selectively activating sensory neurons that convey information about noxious stimuli to the central nervous system. We have used an expression cloning strategy based on calcium influx to isolate a functional cDNA encoding a capsaicin receptor from sensory neurons. This receptor is a non-selective cation channel that is structurally related to members of the TRP family of ion channels. The cloned capsaicin receptor is also activated by increases in temperature.

In the noxious range, suggesting that it functions as a transducer of painful thermal stimuli in vivo." https://www.nature.com/articles/39807

"Capsaicin binds the N-terminus of Hsp90, induces lysosomal degradation of Hsp70, and **enhances the anti-tumor effects of**
17-AAG (Tanespimycin)" Chaitanya A. Patwardhan 1, Vamsi Krishna
Kommalapati 1, Taoufik Llbiyi 1, Digvijay Singh 1, Eyad Alfa 1, Anatolij
Horuzsko 1, Hasan Korkaya 1, Siva Panda 2, Christopher A. Reilly
3, Vladimir Popik 4 & Ahmed Chadli

Hot peppers also construct cRNA, that's right, a cousin of RNA. This is a protein complex, and receptors are where hot peppers connect, **delivering immune strength by empowering DNA.** "cDNA that codes for a specific protein can be transferred to a recipient cell for expression, often bacterial or yeast expression systems." Croy, Ron. "Molecular Genetics II - Genetic Engineering Course (Supplementary notes)." Durham University durham.ac.uk; 20 April 1998. Archived from the original on 24 August 2002. Retrieved 4 February 2015.

People who regularly eat hot peppers are healthy and vibrant. Your diet should have a healthy number of peppers with Curcumin and Terpenes.

Terpenes

Terpenes that are prominent and especially helpful for cancer inhabitation, "Terpenoids composed of "isoprenoid" units constitute one of the largest groups of natural products accounting for more than 40, 000 individual compounds, with several new compounds being discovered every year [42-44]. Most of the terpenoids are of plant origin; however, they are also synthesized by other organisms, such as bacteria and yeast, as part of primary or secondary metabolism. Both CBD oil and THC have terpenes. Terpenoids are a diverse group of compounds found in various plants, and they have a wide range of uses. Here are some examples of terpenoid compounds and their uses:

Limonene: Found in citrus fruits, limonene is used in the production of fragrances, cleaning products, and as a flavoring agent.

Menthol: Derived from peppermint and other mint plants, menthol is commonly used in topical products for its cooling and soothing effects on the skin. It is also used in cough drops and other throat lozenges.

Camphor: Obtained from the camphor tree, camphor is used in traditional medicine for its analgesic and anti-inflammatory properties. It is also used in topical products for its cooling effect and as an insect repellent.

Taxol: Derived from the Pacific yew tree, Taxol is a potent anticancer drug used in the treatment of various types of cancer, including breast, ovarian, and lung cancer.

Artemisinin: Found in Artemisia annua (sweet wormwood), artemisinin is a key component in the treatment of malaria. It is used in combination with other antimalarial drugs to combat drug-resistant strains of malaria.

Geraniol: Found in various plants, including roses and lemongrass, geraniol is used in the production of perfumes, soaps, and other cosmetic products. It also has antimicrobial and insect-repellent properties.

Curcumin: Derived from turmeric, curcumin is a well-known antioxidant and anti-inflammatory compound. It is used as a dietary supplement and is being studied for its potential therapeutic effects in various diseases, including cancer, arthritis, and neurodegenerative disorders.

These are just a few examples of the many terpenoid compounds found in nature and their diverse uses in various industries, including pharmaceuticals, cosmetics, and fragrances. Studying terpenes is important for several reasons. It can lead to the development of new drugs and treatments, improve flavors and fragrances, provide insights into plant-insect interactions and plant defense mechanisms, contribute to sustainable and efficient industrial processes, enhance terpene production in crops, improve plant resistance to pests and diseases, and develop new bio-based products. Additionally, studying terpenes can help in understanding their role in air pollution and climate change, contribute to the knowledge of biodiversity and evolution, and lead to the discovery of new natural products and the development of innovative biotechnological processes.

(A+B) exhibited cytotoxicity, cell killing, against Bel-7402 cells.

It increased apoptosis and decreased the proliferation of the cells.

It caused microtubule disassembly in the cells.

"Table 1

In vitro effects of natural terpenoids on various liver cancer cells. Terpenoids Compounds Cellular Effects Mechanisms Conc. Ref.

Monoterpenes Geraniol Inhibited the growth of HepG2 cells
The cytotoxic effects of (A+B) on Bel-7402 cells are as follows:
↓HMG-CoA reductase 50-400 μmol/L Polo et al[68]

13.

Diterpenes Andrographolide Inhibited the growth of Hep3B cells
↑apoptosis; ↑ MAPKs; ↑pJNK; ↑ERK1/2 3-100 μmol/L
Ji et al[69]

Excisanin A Decreased viability of Hep3B cells ↑apoptosis;
↓pAKT 1-32 μmol/L Deng et al[70]

Gnidimacrin Demonstrated tumor inhibition in PKC βII-transfected
HLE cells ⊥G2; ↓cdc2; ↑p21WAF1/CIP1 0.00001-0.1 μg/mL Yoshida et al[71]

Oridonin Promoted cytotoxicity in HepG2 cells ↑apoptosis; ↑ROS;
↑p53; ↑p38; ↓ΔΨm; ↑cyt. c; ↑caspase-3, -9 10-50 μmol/L Huang et al[72]

Triterpenes Actein Inhibited the growth of p53-positive
HepG2 cells 27 μg/mL [IC50] Einbond et al[75]

Ardisiacrispin (A+B) Exhibited cytotoxicity against Bel-7402 cells ↑apoptosis;
↓proliferation; microtubule disassembly 1-10 μg/mL Li et al[76]

Zerumbone Showed antiproliferative activity against HepG2 cells ↑apoptosis; ↓Bcl-2; ↑Bax
3.45 μg/mL [IC50] Sakinah et al[127]"
https://www.ncbi.nlm.nih.gov/pmc/articles/PMC3182282/

The HepG2 cells are" An immortalized cell line is a population of cells from a multicellular organism which would normally not proliferate indefinitely but, due to mutation, have evaded normal cellular senescence and instead can keep undergoing division." https://en.wikipedia.org/wiki/Immortalised_cell_line This shows some of the scientific evidence as to why 24 states have legalized marijuana medically and most recreationally because of the major role terpenes have in liver health, which is the main immune processing center. Pharmaceuticals have the opposite effect by inhibiting the functions of the liver and lowering immunity. "Terpenes are the compounds responsible for the way most plants smell. They are abundant in cannabis and many other plants, fruits, and herbs. They may play a role in the health and survival of the plants in which they exist. In some cases, smelling specific essential oils, ingesting terpene-rich plants, or spending time in areas with high terpene content may provide some health benefits to humans. Researchers must continue to study terpenes to understand their potential therapeutic and medicinal uses fully."
https://www.medicalnewstoday.com/articles/whathttps://www.medicalnewstoday.com/articles/what-are-terpenesareterpenes#summary

When you eat plants, there is pollination, and terpenes attract pollination, aiding in the immune defense system.

Green Plants

When you look at the earth, it is mainly green plants and blue water where 40,000 thunderstorms are happening every day to fertilize and grow our medicine.

Best plants for nutrition:

1. Spinach
2. Broccoli
3. Carrots
4. Garlic
5. Brussel Sprouts
6. Kale
7. Green peas
8. Swiss Chard
9. Beets
10. Asparagus
11. Red Cabbage
12. Sweet Potatoes
13. Collard Green
14. Cauliflower

https://www.healthline.com/nutrition/14healthiesthttps://www.healthline.com/nutrition/14-healthiest-vegetables-on-earthvegetables-on-earth#TOC_TITLE_HDR_14

Science reports on the tumor-fighting levels of Terpenes. "Despite significant advances in medicine, **liver cancer**, predominantly hepatocellular carcinoma, remains a major cause of death in the United States as well as the rest of the world. As limited treatment options are currently available to patients with liver cancer, novel preventive control and effective therapeutic approaches are reasonable and decisive measures to combat this disease. Several naturally occurring dietary and non-dietary phytochemicals have shown enormous potential in the prevention and treatment of several cancers, especially those of the gastrointestinal tract. Terpenoids, the largest group of phytochemicals, traditionally used for medicinal purposes in India and China, are currently being explored as anticancer agents in clinical trials. Terpenoids (also called "isoprenoids") are secondary metabolites occurring in most organisms, particularly plants. More than 40 000 individual terpenoids are known to exist in nature, with new compounds being discovered every year. Many terpenoids exhibit cytotoxicity against a variety of tumor cells and **cancer preventive** as well as anticancer efficacy in preclinical animal models. This review critically examines the potential role of naturally occurring terpenoids, from diverse origins, in the chemoprevention and treatment of liver tumors."

"Vegetables high in nitrate include (3Trusted Source):

- Celery
- Cress
- Chervil
- Lettuce
- Beetroot
- Spinach
- Arugula"
  https://www.healthline.com/nutrition/howhttps://www.healthline.com/nutrition/how-to-increase-nitric-oxideto-increasenitric-oxide

"The Lamiaceae family is renowned for its terpenoid-based medicinal components, but Leonurus stands out for its alkaloidrich composition. Leonurine, the principal active compound found in Leonurus, has demonstrated promising effects in reducing blood lipids and treating strokes in addition to the traditional medicinal uses of the herb."
https://www.cell.com/molecular-plant/fulltext/S1674https://www.cell.com/molecular-plant/fulltext/S1674-2052(23)00358-1?dgcid=raven_jbs_aip_email2052(23)003581?dgcid=raven_jbs_aip_email

"In addition to these universal physiological, metabolic, and structural functions, many specific terpenoid compounds (commonly in the CIO, C15, and C20 families) serve in communication and defense, for example, as attractants for pollinators and seed dispersers, competitive phytotoxins, antibiotics, and herbivore repellents and toxins (Harborne, 1991). Terpenoids available in relatively large amounts as essential oils, resins, and waxes are important renewable resources and provide a range of commercially useful products, including solvents, flavorings and fragrances, adhesives, coatings, and synthetic intermediates (Zinkel and Russell, 1989; Dawson, 1994). Members of the terpenoid group also include industrially useful polymers (rubber, chicle) and several pharmaceuticals (artemisinin, taxol) and agrochemicals (pyrethrins, azadirachtin)." https://www.ncbi.nlm.nih.gov/pmc/articles/PMC160903/pdf/07 1015.pdf

As you see, a vast number of terpenes interact between pollination and immune therapy for your body. The most interesting thing is the seed attraction terpenes and pollination connections in a

Photo, or light, coded communication. We are so much a part of the flowers and trees, the vegetation, and the solar energy.

Honey is a product of cross-pollination, and Terpenes attract pollen as mentioned above. So, Honey and CBD/THC oil complement each other and gather forces of immune strength. **Flower Power.**

The air we breathe and the plants we eat interact to defend us. This fuel can make possible the Cultivation of Energy increasing to mega-speeds, zapping toxins and flushing them out of your body.

What people tend to think from a lifetime of media brainwashing is that food is not medicine. All pharmaceuticals are scientifically discovered metabolic mathematical codes from your organic functions. This means that some plants and fruits have the same coded information in a very high number, 50 million per transmission from RNA to DNA, pumping with energy. What science has done to overtake the immune system by lowering it and imitating RNA codes, inflating coded levels, and inflaming with side effects from formula manipulation.

There are two pump receptors called the synapse and dendrite that receive and transmit. This information is the blueprint of your body for the genetic and phonetic expression of your entire body.

"Key points

Neurons communicate with one another at junctions called synapses. At a synapse, one neuron sends a message to a target neuron—another cell.

Most synapses are chemical; these synapses communicate using chemical messengers. Other synapses are electrical; in these synapses, ions flow directly between cells.

At a chemical synapse, an action potential triggers the presynaptic neuron to release neurotransmitters. These molecules bind to receptors on the postsynaptic cell and make it more or less likely to fire an action potential."
https://www.khanacademy.org/science/biology/humanbiology/neuron-nervous-system/a/the-synapse

"Neurons communicate at synapses through a process called synaptic transmission. This involves the following steps:

Generation of an action potential in the presynaptic neuron.

Release of neurotransmitters from the presynaptic terminal into the synaptic cleft.

Diffusion of neurotransmitters across the synaptic cleft.

The binding of neurotransmitters to receptors on the postsynaptic neuron.

Generation of postsynaptic potentials in the postsynaptic neuron.

Integration of signals at the axon hillock of the postsynaptic neuron.

Transmission of the signal to the next neuron if an action potential is generated. Neurons communicate at synapses through a process called synaptic transmission. This involves the following steps:

- Generation of an action potential in the presynaptic neuron.
- Release of neurotransmitters from the presynaptic terminal into the synaptic cleft.
- Diffusion of neurotransmitters across the synaptic cleft.
- The binding of neurotransmitters to receptors on the postsynaptic neuron.
- Generation of postsynaptic potentials in the postsynaptic neuron.
- Integration of signals at the axon hillock of the postsynaptic neuron.
- Transmission of the signal to the next neuron if an action potential is generated.

"https://www.khanacademy.org/science/biology/humanbiology/neuron-nervous-system/a/the-synapse

Chemicals and electrical neurotransmitters combine to start the growth of information imprinted on DNA. The Sun, with 118 elements, radiated the earth which, through billions of years, converted from fire to water, where life began using oxygen that breaks down to carbon and then regenerates plants in a respiratory function.

When synthetic, or inactive ingredients in vaccines, permeate the cells, the action of the electrical nerve can go into shock, as in any electromagnetic formation.

DNA is passed to us through our ancestors. Talent is passed on in music, athletics, philosophy, sacred beliefs, science, and other beneficial talents. Since DNA is 1 in 250,000,000 on average people, that is like yours, making you very rare and special. Those who have genetic defects can be overcome through body mechanics that grow with the injury, best through energy cultivation. A blind person or an autistic person can generate genius in many talented ways.

For instance, DNA possesses biological information, and your brain can calculate the information into a language that, when enlightened, discovers this information. Your inner self knows what it takes to be healthy, not blinded by poor choices in diet and medicine from empirical media. You can awaken a vast connected highway of energy that nourishes body and soul. You are now using this to read from light transmissions. A book!

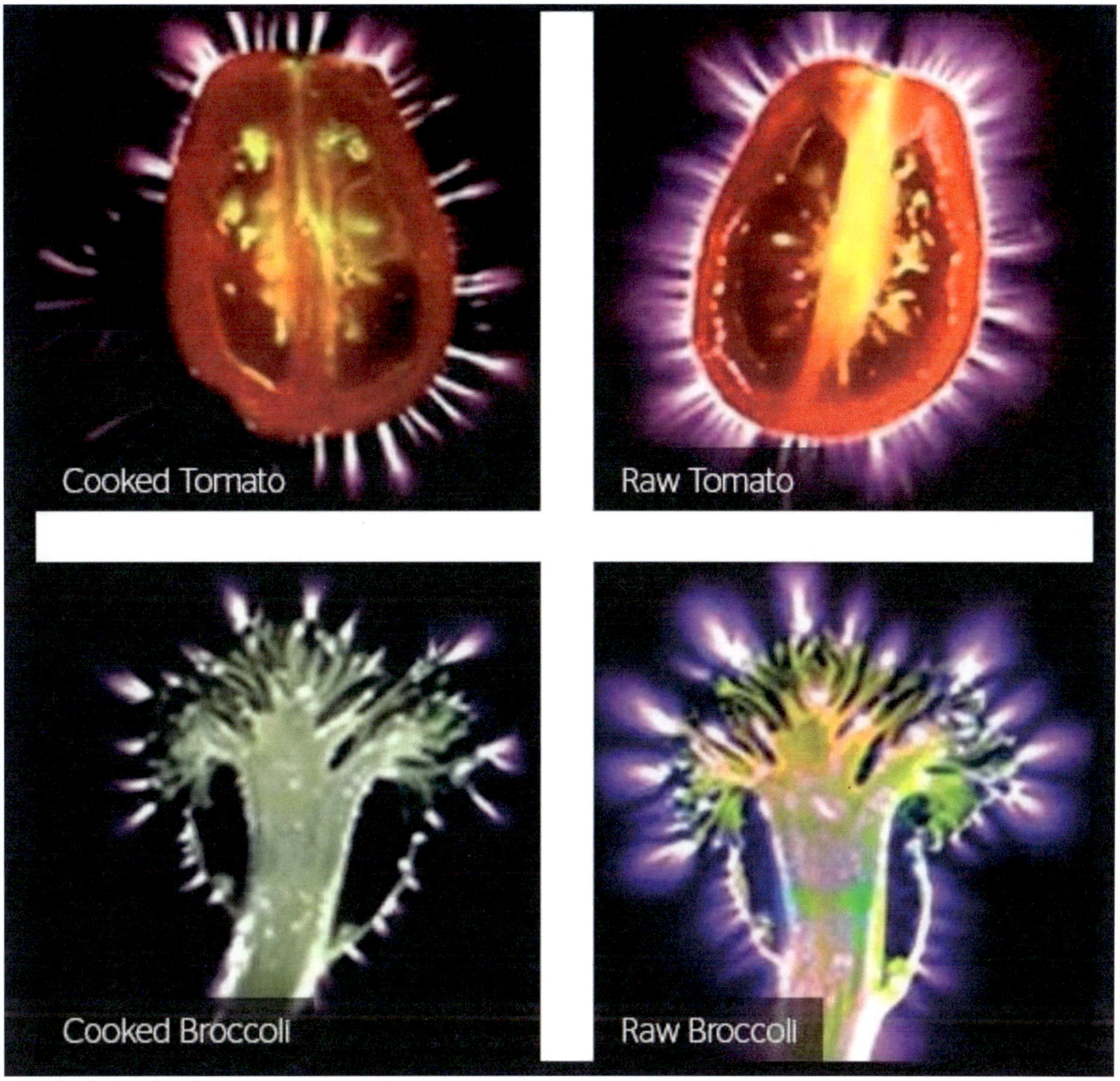

Best Spices for Health:

Ginger

"People have been using ginger for thousands of years as a remedy for nausea and gastrointestinal issues. Today, people primarily use ginger root as a spice or as a supplement to treat digestive problems.

Closely related to turmeric, ginger also contains anti-inflammatory properties Trusted Source, and a very high level of total antioxidants. Only pomegranates and certain types of berries contain more. As an antinausea agent, several controlled studies have proven that ginger is effective as an antiemetic Trusted Source. People can use it for seasickness, motion sickness, and morning sickness.

Some studies Trusted Source have demonstrated that ginger may have anticancer effects. This is largely due to ginger's high antioxidant levels. It can also slow down cells reproducing, cause cells to stop dividing, and stop certain activator proteins and signaling pathways that contribute to cancer. That said, most of these studies took place in a lab, and more human clinical trials must take place.

Cumin

Cumin is a popular cooking spice used for its aromatic effects. Studies Trusted Source show that it is also beneficial for weight loss, cholesterol, stress management, and more. It also has high antioxidant potential Trusted Source.

Research also shows that cumin is an antidiabetic. A group of 80 people took an Ayurvedic formulation containing cumin over a period of 24 weeks, and their postprandial blood sugar was significantly reduced.

Peppermint

An extremely popular herb that is commonly used as a flavoring agent, peppermint is native to Europe and Asia. In these regions, people used it before the advent of modern medicine for its cooling effects, antibacterial properties, and to improve digestive health.

Research shows that, as a holistic remedy, peppermint is also effective in improving cardiovascular (heart) and pulmonary (lung) health by acting as a bronchodilator. Bronchodilators work by widening air passages (bronchioles) in the lungs. By inhaling the smell of peppermint, a person will also increase their nasal air force, in turn supplying more air to the lungs.

In addition, some studies also show that peppermint is an effective muscle relaxer thanks to its cooling menthol compound, which is why menthol is often an active ingredient in ointments and creams that target muscle pain.

Echinacea

Derived from the coneflower plant, echinacea is a supplement that is best known for targeting the immune system and helping prevent colds. While research has yet to prove that the herb can fight viruses, many use echinacea to support the treatment of conditions such as Trusted Source:

- The common cold
- upper respiratory infections
- bronchitis
- influenza
- inflammation
- ear infections.

- vaginitis
- yeast infections.

Echinacea treatments for these illnesses are not proven to be effective.
Most studies show otherwise or are inconclusive at best. There may be a loose correlation Trusted Source between taking echinacea and strengthening the immune system, but much of the evidence for the herb's efficacy is anecdotal.

Cinnamon

Ancient civilizations have used cinnamon since 2,800 BCE for anointing, embalming, and treating ailments. Though not as widely used for its therapeutic properties as it was thousands of years ago, cinnamon still provides myriad health benefits as an antimicrobial, antioxidant, anti-inflammatory, antidiabetic, and anticarcinogenic spice.

A 2015 review Trusted Source shows that cinnamon may reduce blood sugar levels. Its role in regulating glucose in the body has been suggested in many small, randomized control trials, though the results have not been significant enough to prove conclusive.

Cinnamon also has cognitive enhancement effects, and researchers have sought to determine whether the spice could be an effective treatment for Alzheimer's disease. Scientists have studied CEppt (an extract found in cinnamon bark) to ascertain whether it could help prevent the symptoms of Alzheimer's from progressing. When mice ate this extract, it greatly improved their cognitive ability, Trusted Source.

Chili powder

2015 research Trusted Source suggests that capsaicin, the phytochemical that makes chili powder spicy, may play an important role in regulating heart and metabolic health.

When people consume chili powder, the spice triggers beneficial protein changes in the body that are conducive to weight loss. But researchers do not fully understand the mechanism of how this works. People who participated in a study Trusted Source where they consumed moderate amounts of chili over a 12-week period experienced weight loss results, which was triggered by chili's impact on the body to better control insulin, among other therapeutic effects. In another study, Trusted Source, researchers observed that regular consumption of chili also significantly reduced abdominal adipose tissue (fat) levels and reduced appetite and energy intake.

When it comes to cardiovascular benefits, recent research provided by the American Heart Association found that those who regularly consume chili powder may reduce their risk for developing heart disease mortality by 26%, Trusted Source. Additionally, frequent chili consumption correlates with a 25% reduction in mortality from any cause and 23% fewer cancer deaths.

Thanks to its anti-inflammatory properties, chili powder could also be an effective Trusted Source for supporting arthritis treatments, as well as for alleviating muscle and joint inflammation.

Parsley

Parsley is an herb that originated from the Mediterranean region. Many have used it through the years for both culinary flavoring and therapeutic treatment for a range of health conditions, including high blood pressure and allergies.

This herb is full of antioxidants, carotenoids, and other beneficial vitamins that support a healthy body and immune system. Among these is vitamin K, an essential nutrient for bone health.

Oregano

Another herb that is prevalent in the Mediterranean diet is oregano, which many use not only as a flavor enhancer for food but also as an aromatic oil and supplement. The herb's antioxidants contribute to its powerful taste and smell, and may also provide benefits, Trusted Source such as:

- strengthening the immune system against infections
- reducing inflammation
- regulating blood sugar
- improving insulin resistance
- alleviating urinary tract symptoms and menstrual cramps
- fighting cancer

Oregano, as a standalone herb, cannot singlehandedly produce these results. But people may use it for its antioxidative properties that are beneficial for many health conditions.

Cardamom

Cardamom is yet another spice that provides a wide range of potential health benefits. Coming from the seeds of plants that belong to the ginger family, people often consume cardamom in teas, such as chai tea, as well as some coffee, desserts, and even savory dishes.

Some of the conditions that research Trusted Source shows cardamom may treat are:

- constipation
- colic
- diarrhea
- dyspepsia
- vomiting
- headache
- epilepsy

- cardiovascular disease.

The spice's healing benefits come primarily from a combination of its volatile oils, fixed oils, phenolic acids, and sterols. In particular, the volatile oils present in cardamom seeds are analgesic, anti-inflammatory, antimicrobial and antispasmodic.

Animal studies found cardamom may also help with obesity or high cholesterol. When researchers implemented cardamom supplementation on obese rats, the spice lowered their total cholesterol Trusted Source and triglyceride levels." https://www.medicalnewstoday.com/articles/healthy-herbshttps://www.medicalnewstoday.com/articles/healthy-herbs-and-spicesandspices#cardamom

Chili peppers, in general, add not only spice but are beneficial in healing, as explained.

Incorporating Jalapenos, Poblanos, and Serranoes, to mention a few, will enhance flavor by roasting and included in beans. This meal with pinto beans is filled with nutrients, including proteins and capsaicin.

Beans have numerous health benefits, including being a rich source of plant-based protein, high in dietary fiber for digestion and blood sugar control, packed with essential nutrients, promoting heart health, aiding in weight management, supporting gut health, providing antioxidants, contributing to bone health, and being versatile in cooking. They are also nutrient-rich, promote digestive health, may help prevent cancer, and support energy and brain function.

The AMA and FDA OK opium addiction, snake poison chemotherapy, side effects, gastrointestinal infection, blocking protein pumps, shutting down the liver functions, thins the blood, vaccinates, which is probable cause muscular sclerosis and lowers immunity energy, and does not support good food for medicine. The FDA allows 5000 chemicals, and cancer caused by pesticides and preservatives that are micro-solutions of metal, processed sugars crystallize into salt that causes corrosiveness by all physical science.

However, energy from nature can precisely heal without side effects, but can improve the energy. It is made from sunshine and earth and develops proteins that generate energy and improve immunity. It's not medicine, it's biologically energy-efficient food. Plants convert protein energy, where mass becomes energy, moving from Einstein's formula for energy through metabolic pathways. **One point to remember all food and supplements are the best sources when organic**. Inorganic uses inactive ingredients that are metabolic residual, not digested or broken down properly, which generate side effects.

Terpenes also have attracted pollination, as previously discussed. Honey is a source of pollination and combines with honeycomb. When you consume honey and terpenes in food, such as CBD, and THC, the pollination will be powerful in your diet. The elasticity of hemp and marijuana stretches more than any other element. The flash point where it breaks down survives longer and grows like a weed. "Phenotypic plasticity is the potential of a genotype to form different phenotypes in contrasting environments. Phenotypic plasticity is always present among plant leaves due to the modularity of design, such that individual leaves can acclimate to their own environment. Plasticity differs among genotypes, populations, and species, and as a result, plants vary in their capacity to reach the optimum phenotype and maximize fitness under different environmental conditions." https://onlinelibrary.wiley.com/doi/10.1002/9781119312994.apr 0714

"Most powerful herbs and sources of health benefits.

- THC/ Terpenes
- CBD/ Terpenes
- Ginko
- Turmeric
- Honey
- Chili Peppers/ Curcumin
- Lavender
- Grape Seed
- Chamomile
- Echinacea" https://www.healthline.com/health/most-powerfulmedicinalplants#echinacea

These plants and elements from the ground and Solar Radiation will explode in formula but will be cohesive in your body, containing the explosion from Hydrogen, Nitrogen, and Oxygen. This will not adhere to your body's metabolic function in your cellular pathway. These elements are used to make nuclear weapons. You have immense power contained in the atmosphere for detoxing.

What adheres is the metallic solution in your body from Vaccines.
Preservatives, Pharmaceuticals, and Pesticides, the VPPPX factor. This is defined as inactive ingredients on the label of the products. They will collect and start blocking your energy and metabolic pathways until you eat or consume nutrition free of the inactive ingredients, detoxing with the Phyto-Photon of immune defense forces.

"Ingredients keep vaccines safe and long-lasting.

Some ingredients help make sure a vaccine continues to work like it's supposed to and that it stays free of outside germs and bacteria. For example:

Preservatives, like thimerosal, protect the vaccine from outside bacteria or fungi. Today, preservatives are usually only used in vials (containers) of vaccines that have more than 1 dose. That's because every time an individual dose is taken from the vial, it's possible for harmful germs to get inside. Most vaccines are also available in single-dose vials and do not have preservatives in them.

Stabilizers, like sugar or gelatin, help the active ingredients in vaccines continue to work while the vaccine is made, stored, and moved. Stabilizers keep the active ingredients in vaccines from changing because of something like a shift in temperature where the vaccine is being stored."
https://www.hhs.gov/immunization/basics/vaccineingredients/in dex.html

Keeping them long-lasting is so they can preserve the vaccines, and how about you? Micro-solution of metal that corrodes the body, and if you eat processed sugar, that's also salt, cook the corn syrup until it becomes crystallized into salt. Then it helps to corrode your body when the metals develop and rise to competition for survival, and the salts converge, then corrosion exists in the inactive ingredient in your body and adheres, as nature does to survive. Attaches to the walls of metabolic pathways of veins and arteries.

"What Is High Fructose Corn Syrup?

You've probably heard of high fructose corn syrup, but what is it? Put simply, it's a sweetener derived from cornstarch.

Cornstarch is composed of a chain of glucose (simple sugar) molecules joined together. Corn syrup, which is basically 100 percent glucose, comes from the breakdown of cornstarch into individual glucose molecules.

In order to create **high fructose corn syrup**, enzymes must be added to corn syrup to change some of the glucose into another simple sugar called fructose. The enzymes, alpha-amylase and glucoamylase, used in HFCS processing have been genetically modified to improve their heat stability for the production of HFCS.

According to the U.S. Food and Drug Administration, most high fructose corn syrups contain either 42 percent or 55 percent fructose. The rest of the HFCS is glucose and water.

HFCS 42 is typically what's used in cereals, processed foods, baked goods, and some beverages. HFCS 55 is used mainly in soft drinks. However, some HFCS contains up to 90 percent fructose." https://draxe.com/nutrition/high-fructose-corn-syrup-dangers/ Now they use **Sucrose,** which is as bad." Sucrose, or table sugar, may add sweetness to your favorite treats, but when it comes to your health, it's not so sweet. Weight gain, poor blood sugar control, and heart disease are just a few of the dangers of regularly including sucrose in your diet."
https://www.livestrong.com/article/205180-what-are-thedangers-of-sucrose/

It simply is all about profit, and to hell with your health. Zero sugar uses Sucrose!

As discussed, liver cancer is one of the top cancers in the population. "In summary, while liver cancer is a significant health issue, it may not be the absolute top cancer disease in terms of incidence. However, its impact on mortality remains substantial. Early detection, prevention, and management are crucial for addressing this disease effectively. If you or someone you know exhibits symptoms related to liver cancer, seeking medical attention promptly is essential."
https://www.bing.com/search?q=is%20liver%20cancer%20the%20top%20cancer%20disease&qs=SYC&showconv=1&sendquery=1&FORM=ASCHT2&sp=1&lq=0&sm=csrmain

"The liver detoxifies various metabolites, synthesizes proteins, and produces biochemicals necessary for digestion. Its other roles include the regulation of glycogen storage, decomposition of red blood cells, and the production of hormones. It is also the largest internal organ in the body.

The liver has many important functions. If it's not working well, it can lead to problems throughout the body.

Here are a few examples of the health complications you may experience if you develop liver damage from medication use:

Filtration: The liver removes unsafe substances (toxins) from the body. Toxins are either passed out of the body in urine and feces or broken down into substances that are safer. If the liver is not working well, these toxins can build up.

Metabolism: The liver is part of many metabolic processes that help the body process energy. If a person's liver isn't working, these processes won't work well.

Blood clotting: The liver also helps blood cells clump together to stop bleeding (blood clotting). A person who doesn't have a
healthy liver is at risk for bleeding problems." https://www.verywellhealth.com/drug-induced-liver-disease1943023#:~:text=Certain%20oral%20and%20topical%20antibiotics%20are%20linked%20to,3%20Amoxicillin-clavulanate%204%20Azithromycin%205%20Tetracycline%206%2 0Clindamycin

Medications and pharmaceuticals are damaging to the liver along with alcohol, Tylenol, and processed sugars. Statins block liver functions, which is energy draining." Statins can have a lot of side effects, including muscle pain, digestive problems, and liver damage." NonBrand&gclid=d22a57aefb7d1ac661af31a73e6f7ffa&gclsrc=3p.ds&msclkid=d22a57aefb7d1ac661af31a73e6f7ffa&utm_source=bing&utm_medium=cpc&utm_campaign=Health-HeartHealth-NonBrand-Phrase&utm_term=side%20effects%20of%20cholesterol%20medications&utm_content=ConditionsTreatments-StatinMyths

By detoxing and using respiration or breath control, the body can loosen the adherence to immune functions from the consumption of plant power cleaning.

When this happens, the detoxing increases energy with megaspeed cleaning power. The energy cultivated by energy-giving food helps you eat your way to health.

Your computer has all kinds of virus protection; some are better than others because they do not let inactive algorithms corrupt computation. Your organic computer needs active ingredients that fuel immune functions, detoxing the body for efficient energy production.

A good diet and exercise are well known for the activity being in shape, but the FDA has allowed synthetic chemicals to be produced, making the feat harder. Organic foods are free of toxins, and certain herbs and plants have higher levels of detoxing; Respiration allows combustion for carbon-reducing unhealthy proteins. This is accomplished in a chemical process utilizing power elements of Oxygen Hydrogen, and Nitrogen in a controlled method of ionization equilibrium, cohesively executing energy, reducing toxins to carbon, regenerating the tissue's immune strength. As you have seen, energy is what makes you tick, and I have demonstrated with full documentation what science knows, the government knows but is not advertised because the profits of the food and drugs are greater than the morality of healing.

The answer is, are you willing to take a step to begin a healthy lifestyle?

Dieting and exercise are the best and most steady plans for healing.

The first step is the hardest to take; each step is then energized as you fuel with nutrients, the essential elements for formula production.

To make it simple, eat a steady diet of organic nuts, berries ang green plants in as wide a variety as you can.

Eat less red meat because it is acidic for digesting and tougher to digest. White meat and fish with skin on are beneficial.

I have given you science to show how good nutrition is a reality, and how processed and mass-produced food and medicine all have bad side effects.

Hydration

This is a simple rule: weigh yourself and divide by 2 that equals the ounces you need. A 190 lb. man is then supposed to drink 85 ounces of water daily.

Water that has low acid levels and does not use fluoride to control bacteria is the rule for bottled water. Google water for acidic levels that are low and have higher alkaline levels. Many of the big names will use acid to kill bacteria, and better ones use alkaline to neutralize bacteria that are unhealthy and contaminated.

Check the best water sources!

https://www.bestproductsreviews.com/bottled-water-brands-of-2022?targetid=dat-2336187331767387:loc-190&matchtype=b&device=c&campaignid=530764315&creative=&adgroupid=1350202779891234&feeditemid=&loc_physical_ms=97737&loc_interest_ms=&network=o&devicemodel=&placement=&keyword=$bottle&target=&adposition=&trackid=us_all_top_2_bing&mId=H149003CQN&msclkid=179217fe4ccc179c956770b014b22620

Reverse Osmosis is a good investment for healthy water. "Reverse osmosis is a multi-stage water treatment process that removes contaminants from unfiltered water, or feed water, when pressure forces it through a semipermeable membrane. Residential reverse osmosis systems utilize at least three stages of treatment to reduce levels of virtually all contaminant types. In the final stage, the RO membrane, water flows from the more concentrated side (more contaminants) of the membrane to the less concentrated side (fewer contaminants) to provide clean drinking water. The fresh water produced is called permeate. The concentrated water left over is called the waste or brine."

https://www.freshwatersystems.com/blogs/blog/what-is-reverseosmosis

Reverse osmosis removes:

Fluoride

Salt

Sediment

Chlorine

Arsenic

VOCs

PFAS

Microplastics

Herbicides and pesticides

Heavy metals

Many other contaminants"
https://www.freshwatersystems.com/blogs/blog/what-is-reverseosmosis

Nature uses this naturally with rain and sun, vast oceans, creeks, rivers, and lakes! We need this flushing to detox, and along with nutrients, we clean our body!

Combining nutrients and water, we have made it possible for the growth of the body, and an active lifestyle is also needed. The next chapter will concentrate on exercises designed to stretch every muscle!

Full Compass Stretching

Glenn Moran and Andrea Teja perform Yoga positions, called Asanas, to work every muscle in the body through full directional stretches. My book Yoga Light shows over 100 positions in eight workout plans. Available on Amazon and other book sellers. The ones here are excellent and, combined with Breath Control, Pranayama will elevate fitness and your range of flexibility.

Warming up to perform other positions, Asanas, is to loosen tendons and muscles and begin working out the Cat and Cow with variations on this theme are excellent. Stimulating muscle groups allows exercise to distribute blood and, with nutrition consumed, produces muscle and tendon development in every part of the body engaged with proper full compass, or multi-directional exercising. Also, the nutrients give you viscosity fluids called Synovial Fluid. With exercising and proper diet, these fluids "grease the joints." Many people, without understanding, get knee surgery when they can diet and exercise. People say it's bone on bone, but it is more accurate that you are not eating enough organic fruits and vegetables.

"Synovial Fluid

Synovial fluid (SF) exists as a viscous lubricant in flexible joints such as the knee and elbow."
https://www.sciencedirect.com/topics/medicine-anddentistry/synovial-fluid#:~:text=Synovial%20fluid%20(SF)%20exists%20as,than%20sufficient%20for%20alcohol%20analysis.

"Synovial Fluid

Synovial fluid (SF), present in very small quantities in normal synovial joints, has two functions: lubrication and nutrition.39,40 Normal fluid is clear and pale yellow; SF is a combination of a filtrate of plasma, which enters the joint space from the subsynovial capillaries, and hyaluronic acid, which is secreted by the synoviocytes. Hyaluronic acid provides the high viscosity of normal SF and, with water, its lubricating properties.41 Concentrations of small molecules (electrolytes, glucose) are similar to those in plasma, but larger molecules (e.g., complement components) are present in low concentrations relative to plasma unless an inflammatory state alters vasopermeability. Notably absent from SF are elements of the coagulation pathway (fibrinogen, prothrombin, factors V and VII, tissue thromboplastin, and antithrombin).42 As a result, normal SF is resistant to clotting. There appears to be free exchange of small molecules between SF of the joint space and water bound to collagen & cartilage
https://www.sciencedirect.com/science/article/abs/pii/B978032324145 8000028

The body, with proper nutrition, will convert to energy that will build muscle and provide the necessary grease for your joints. Exercise will stimulate this energizing healing.

Cat and Cow

"Cat-Cow is a gentle flow between two poses that warms the body and brings flexibility to the spine. It stretches the back torso and neck, softly stimulates and strengthens the abdominal organs. It also opens the chest, encouraging the breath to become slow and deep. The spinal movement of the two poses stimulates the kidneys and adrenal glands. Coordinating this movement with your breathing relieves stress and calms the mind.

This sequence also helps to develop postural awareness and balance throughout the body. It brings the spine into correct alignment and can help prevent back pain when practiced regularly.

Cautions

Those with neck injuries should keep their head in line with the torso, not dropping it forward or back. Pregnant women and those with back injuries should only perform Cow Pose, bringing the spine back to neutral between poses — do not let the belly drop between repetitions, as this can strain the lower back. Always work within your own range of limits and abilities. If you have any medical concerns, talk with your doctor before practicing yoga.

Instructions:

1. Start on your hands and knees with your wrists directly under your shoulders, and your knees directly under your hips. Point your fingertips to the top of your mat. Place your shins and knees hip-width apart. Center your head in a neutral position and soften your gaze downward.
2. Begin by moving into Cow Pose: Inhale as you drop your belly towards the mat. Lift your chin and chest and gaze up toward the ceiling.
3. Broaden your shoulder blades and draw your shoulders away from your ears.
4. Next, move into Cat Pose: As you exhale, draw your belly to your spine and round your back toward the ceiling. The pose should look like a cat stretching its back.
5. Release the crown of your head toward the floor, but don't force your chin to your chest.
6. Inhale, come back into Cow Pose, and then exhale as you return to Cat Pose.
7. Repeat 5-20 times and then rest by sitting back on your heels with your torso upright." https://www.theramotion.com/news/did-you-know-thecat-cow-yoga-stretch-has-many-benefits/

Cat

Cow

Cat Cow with Leg Extended

Adding variations to the Cat-Cow to increase stretching of the back and torso is the Leg Lift.

Cat-Cow Leg Lift. Bring the leg toward the chest and extend again in fluid movements 4 times with each leg.

Cat-Cow Leg Grasps. Reach back, grasp leg, hold for 5-20 seconds.

Extend arm and leg, and then place arm behind back and upward, holding it for 5-20 seconds.

Place your arm inside your leg and raise your arm upward.

These positions stretch muscles in many directions, loosening muscle groups for other Asanas. Warming up is important to help your whole body so you are relaxed and ready for a Sun Salutation.

Extended Child and Cobra

The lower back is a junction of the spine with over 75,000 nerves. "The cauda equina is a group of nerves and nerve roots stemming from the distal end of the spinal cord, typically levels L1-L5 and contains axons of nerves that give both motor and sensory innervation to the legs, bladder, anus, and perineum."

Many people suffer from a lack of stretching and are sedentary level. This exercise should be done with care. The back and knees get a good stretch.

Remember that all yoga is performed short of the pain threshold. In other words, do half the length of stretching if there are back problems.

Extended Child and Cobra Warm Up.

Extended Child Description

"From a kneeling position, the toes and knees are together, with most of the weight of the body resting on the heels of the feet. The arms extend to the front with the fingers spread wide. The forehead rests softly onto the earth. The gaze is down and inward."

Benefits

"Gently stretches the hips, thighs, and ankles. Calms the brain and helps relieve stress and fatigue. Relieves back and neck pain when done with the head and torso supported."

https://www.sciencedirect.com/science/article/abs/pii/B978012385157 4011283

The Cobra is followed in movement by the Extended Child. This rotates the lower back, giving benefit to a stronger back. "The king cobra, revered in Indian myths, can glide forward while lifting

the upper third of its body upright. When you practice Cobra Pose, try to emulate this animal's powerful yet fluid motion when you practice. Imagine your legs as the snake's tail, reaching long behind you as you curve your spine to lift your chest majestically."
https://www.yogajournal.com/poses/cobra-pose-2/

Over time, as the Extended Child-Cobra movements bend the torso, you will stimulate the spine and lower back, mega-nerves extending from the lower back, spinal fluids from nutrients and to the lower nerve extremities, relieving back pain radiating in the spine and legs. "The cauda equina is comprised of lumbosacral nerve roots that descend from the spinal cord, which ends at about the L1 vertebral level in an adult, to exit/enter the spinal column through intervertebral foramina in the lumbosacral portion of the spinal column. Pathological processes that impact the cauda equina produce a clinical picture of lumbosacral polyradiculopathies with potential motor weakness and sensory loss in the lower extremities and often bladder and bowel dysfunction. Depending on the nature of the lesion, there also may be low back pain and radiating pain into one or both lower extremities." https://www.yogajournal.com/poses/cobra-pose-2/

Sun Salutation

1. Urdhva Hastasana (Upward Salute)

Stand with your feet together and your arms at your side. Distribute your weight evenly between your feet and turn your palms to face forward in Tadasana. Let yourself come to a slow, steady breath. Find your own center as you bring your palms together at the center of your chest in Anjali Mudra (also known as Salutation Seal or, quite simply, prayer hands)

2. Uttanasana (Standing Forward Bend)

Inhale and sweep your arms straight out to the sides and up alongside your ears into Urdhva Hastasana. Your palms should face one another, and you can bring them to touch if desired. Reach your heart and arms toward the sun. If you like, you can gaze slightly upward and even take a slight backend here, lifting your chest and leaning your upper body slightly back.

3. Ardah Uttanasana (Half Standing Forward Bend)

Exhale and bend forward at your hips, bringing your chest to your thighs. Keep your legs strong by drawing your knees toward your hips. Let your shoulders and neck relax. Your hands can rest on your shins, ankles, or blocks in Uttanasana.

4. Anjaneyasana (Low Lunge)

At the end of your exhale and just before you inhale, step your left foot forward between your hands into Anjaneyasana. Quickly bring your right foot forward next to your left. If you're doing more than a single round of Surya Namaskar A, alternate which foot you step forward first with each round.

5. Plank Pose

Exhale and step your feet back into Plank Pose, which is essentially the top of a push-up. Align your body so your shoulders are directly over your wrists, your palms are flat on the mat, and your feet are hip distance apart. (Lower your knees to the mat if you ache in your lower back, shoulders, or arms.) Gaze down and slightly forward, keeping the back of your neck long. Inhale and lengthen through your back.

6. Cobra Urdhva Mukha Svanasana (Upward-Facing Dog Pose)

Inhale and draw your chest forward and up as you roll forward over your toes onto the tops of your feet into Urdhva Mukha Svanasana. Press the tops of your toes into the mat and keep your legs strong, but relax your gluteal muscles. Pull your shoulders back and broaden through your collarbones. Gaze straight ahead or slightly up. (Instead of rolling over your toes, you can lift and place your feet, one at a time, on the mat. If this backbend feels too intense, instead come into cobra.

Now ascend with Low Lunge on the opposite foot, then perform Half S forward Standing Forward Bend. Then Standing Forward Bend, Upward Salute. It's a reverse of the starting positions of the Asanas.

Warrior Workout

This workout is a great way to invigorate your back. Bending the back in several directions through various movements and holding for 1020 seconds will strengthen the back and knees.

Each movement can start with small extensions according to the present levels of ability to stretch. With practice, increasing stretching will develop. Breathing should be done deeply while doing exercise.

- Extend arms forward and backward, with practice knee is directly over the ankle, the back foot is plated twisting horizontally in direct line from the front foot. Hold for 10-20 seconds.

Extend both arms forward, holding for 10-20 seconds. When you first do this, you may shake. Over time, it will be steady.

- Extend your arm angularly above your head and reach down with the alternate arm, touching the floor inside of your ankle. Hold for 10-20 seconds. Make sure your feet are in line.

- Repeat this on both sides. Remember to gradually increase the level of stretching. While doing the exercise, stop on the threshold of pain and hold where it is comfortable.

Straighten the Knee out and perform Triangular Asana, hold for 10 -20 seconds, and relax.

Repeat the entire Warrior workout on the opposite side.

Beginners, remember your range of motion will increase with practice. Do not try more than you are capable of doing. Stretch gently each workout till you improve your range of motion.

When stretching, you will feel the extent of your range, hold, and breathe deeply. Then try to extend further gently. This will help you increase your range.

Deep breathing also helps the exerciser to calmly execute the exercise.

Pranayama-Breath Control

Yogis say that Breath Control is the most Vital for Yoga and your breathing raises energy to intellectual or spiritual, or grasped by the understanding.

The following exercises give daily bread, 24-7. Improves thoracic chest expansion, improves energy for explosiveness, cell dynamics, and carbon reduction for exhaust. Because the metabolic energy fueled by photons' organic formula overpowers with supercharged explosiveness, neutralizing cancer cells for hydrocarbon trash in your blood paths and finds several exhaust points, exhalation dispels toxins.

The overpowering comes from Hydrogen and Nitrogen. They make bombs by breaking lasers to ignite diving cells at mega speeds, increasing blasting power to carbonize the people in Hiroshima in a 60-mile radius. The body controls this with a formula released slowly to build immunity and restore healthy proteins.

The clouds for thunderstorms are 40,000 a day. This is from ionization and releasing bolts 35 times as hot as the sun. This then fuses Nitrogen and Oxygen, and when you breathe through the nose becomes Nitric-Oxygen that lasers toxins out of the body like a Star Ship, neutralizing cells and landing with precision in a renewed frontier. This is a cellular picture imaging for Macrophage, a protein structure with sub-units formulated by White Blood cells. It attacks unhealthy proteins, growing dangerously. If your diet is high in acid, it is toxic and formulates the protein to attack with subunits to survive. Only with an army of highly sophisticated, complex, energized cells are armed to 20% in your attack from the disease. At 80%, it is invaded, and you need alkaline PH levels under 20%, and eat only organics at least 80%. To counter acid buildup.

"Macrophages are present essentially in all tissues, beginning with embryonic development and, in addition to their role in host defense and in the clearance of apoptotic cells, are being increasingly recognized for their trophic function and role in regeneration." Ovchinnikov DA (September 2008). "Macrophages in the embryo and beyond much more than just giant phagocytes."

This is a virus structure radiating outwards, killing cells to grow, the cell killing or cytotoxic, to survive from toxic fuel. Processed sugar creates fat cells, and natural organic sugar glycol creates energy for your body, represented by the macrophages. "A type of white blood cell that surrounds and kills microorganisms, removes dead cells, and stimulates the action of other immune system cells." The disease is a spiked protein illustrated below.

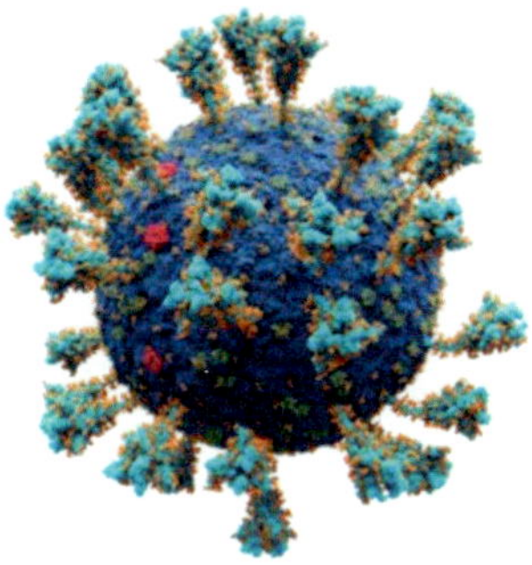

Cytopathic effect or cytopathogenic effect (abbreviated CPE) refers to structural changes in host cells that are caused by viral invasion. The infecting virus causes lysis of the host cell, or when the cell dies without lysis due to an inability to replicate.

Macrophage formulates in four days, from organic nutrients, and attack by wrapping around, consuming virus spikes, and neutralizing by cytotoxic, cell-killing. The healthy proteins planted seeds and were fertilized by your breath with nitric-oxygen free radicals, detoxed the viral invasion, opening pathways for organic forces of energy. “Macrophages are professional phagocytes and are highly specialized in the removal of dying or dead cells and cellular debris.” Eming SA, Krieg T, Davidson JM (March 2007). "Inflammation in wound repair: molecular and cellular mechanisms.” The Journal of Investigative Dermatology. 127 (3): 514–25. doi:10.1038/sj.jid.5700701. PMID 17299434

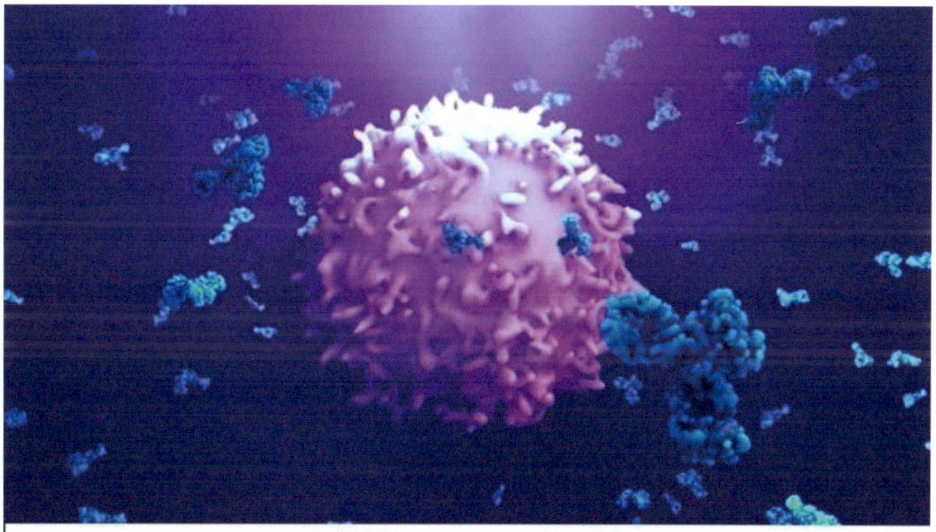

Pranayama is vital for the delivery and exhausting of Volatile Organic Compounds from the debris of toxins cleaned by macrophage function.

“Volatile organic compounds (VOCs) are released within the human organism as a result of normal metabolic activity or due to pathological disorders. They enter the bloodstream and are eventually metabolized or excreted via exhalation, skin emission, urine, etc.” Inzet SpiroNose stappen dichterbij gekomen.” Longfonds. Feb 26, 2020. Retrieved Aug 14, 2020.

"A hydrogen breath test (or HBT) is used as a diagnostic tool for small intestine bacterial overgrowth and carbohydrate malabsorption, such as lactose, fructose, and sorbitol malabsorption." Eisenmann, A.; Amann, A.; Said, M.; Datta, B.; Ledochowski, M. (2008). "Implementation and interpretation of hydrogen breath tests." Journal of Breath Research. 2 (4): 1–9.
Bibcode:2008JBR.....2d6002E. doi:10.1088/1752-7155/2/4/046002. PMID 21386189. S2CID 31706721.

Identification and quantification of potential disease biomarkers can be seen as the driving force for the analysis of exhaled breath.

What is being said is the importance of breathing." VOCs play an important role in communication between animals and plants, e.g., attractants for pollinators, protection from predation,[4] and even inter-plant interactions." Pichersky, Eran; Gershenzon, Jonathan (2002). "The formation and function of plant volatiles: Perfumes for pollinator attraction and defense." Current Opinion in Plant Biology.
5 (3): 237–243. doi:10.1016/S1369-5266(02)00251-0. PMID 11960742.

Wow, how do you like breathing now? Holding your breath at various times will maximize the forces of nitric oxygen to keep working. After 3 months, the body will begin to freshen oxygen and fertilize nutrients to be sub-unit forces and special forces immune defense, increasing lung cell growth.

I had sinus problems for 40 years, took antihistamines all the time, and Prana cured it for the last 15 years. I take no pills and have a clear sinus. Constant breathing exercises that I will share with you. A friend of mine relieved 40 years of asthmatic disease to the amazement of the doctors.

The first thing you must learn is Diaphragm Breathing. Many people have what I call a short breath, filling the top of their lungs first. Maximizing the Vital Force of nitric-oxygen inhalation and all atmospheric organic levels is to fill the bottom of the lungs first and fill it to the top last. This is done as follows:

Diaphragm Breathing

The breathing control is to expand the lower lung by pushing the diaphragm out to fill the lungs from the bottom to the top and then pushing the diaphragm towards the spine to empty the lungs. When you fill a pitcher with water, the bottom is filled first. The diaphragm is a muscle, and this control will help strengthen the belly. This will increase the oxygen throughout your body dramatically. The air we breathe is the same energy that inspires all physical matter and releases tension. During the Yoga postures, or Asanas, breath control is done rhythmically.

1. Inhale, pushing the diaphragm away from the spine
2. Exhale, pushing the diaphragm muscle towards the spine
3. Place your hand on your stomach and notice the breath direction

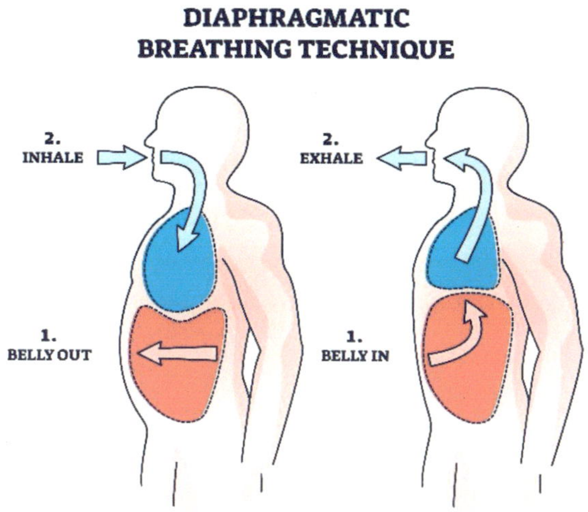

Practice this, and it becomes an instinctual function of the body that will increase dynamically over time as you inhale and exhale, a respiration process of detoxing production, filling the lungs. The average person gets around 50% less oxygen when breathing from the top of the lung.

“Aging is proportional to respiratory rate (excluding exceptions), e.g.,

Mouse: Respiratory rate = 60-230/min and Life span = 1.5-3.0 years

Rabbit: Respiratory rate = 30-60/min and Life span = 5.0-6.0 years

Monkey: Respiratory rate = 30-50/min and Life span = 20–30 years Human: Respiratory rate = 12-16/min and Life span = 70–80 years

Whales: Respiratory rate = 3-5/min and Life span = more than 100 years” https://www.quora.com/What-factors-decide-the-maximumlifespan-of-an-animal-including-humans

Breathing right will treat your body right. We have an immune system that needs atmospheric gases and exhaling the toxic gases from the combustion of functioning immune systems. Maximizing this is important for your health and welfare.

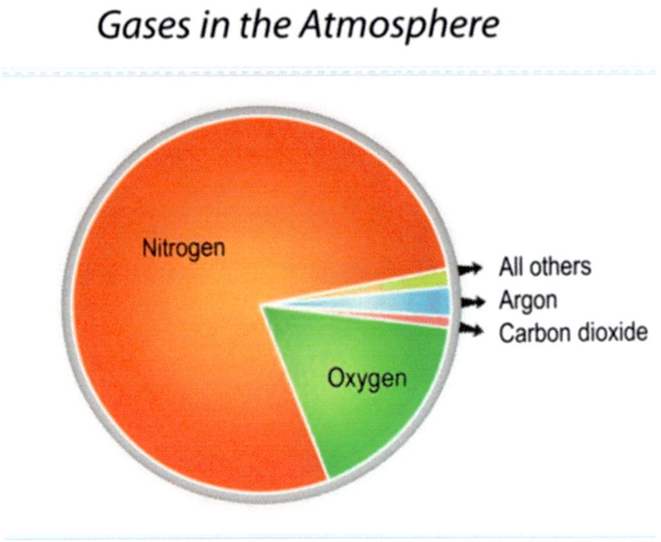

Wearing masks is very unhealthy. Because atmospheric gases are essential for immune processes. Exhalation is a byproduct of detoxing and needs to be exhaled out of the body.

"TABLE 1. Classes of Volatile Organic Compounds in the Breath

Class of VOC Example

Example of Mechanism of Production

Saturated hydrocarbons Ethane, pentane, aldehydes. Lipid peroxidation of fatty acid components of cell membranes–triggered by reactive oxygen species.

Unsaturated hydrocarbons: Isoprene, Mevalonic pathway of cholesterol synthesis.

Oxygen-containing Acetone Decarboxylation of acetoacetate from lipolysis or lipid peroxidation.

Sulfur-containing Ethyl mercaptane, dimethylsulfide

Incomplete metabolism of methionine

Nitrogen-containing Dimethylamine, ammonia Elevated in liver impairment and uremia." https://www.sciencedirect.com/science/article/pii/S155608641533 3529#:~:text=Volatile%20organic%20compounds%20are%20able,c ompounds%20different%20than%20normal%20cells.

Nothing better than fresh air and practicing Breath Control to maximize respiration with efficient inhalation and exhalation. Also called inspiration and expiration, the respiration of interplanetary connection. of the atmosphere. When you breathe, it gives you an immune function aid, and when you exhale, it sends the byproducts of Volatile Organic Compounds into the atmosphere. This is where it is broken into carbons that are regenerated into carbohydrates, fruits, and vegetables, for consumption and production of immune forces of cellular energy.

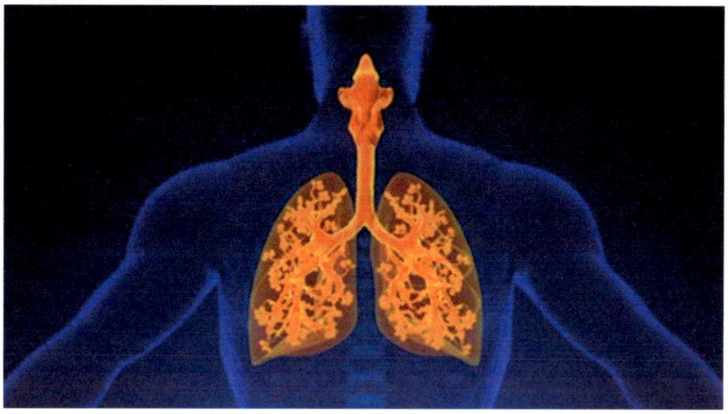

Alternate Sinus

This is the alternate breath with retention, holding breath (Kumbhaka):

1. Clear your nose. Put your right hand up to your nose. Place your index and middle fingers on your forehead, with the thumb and ring fingers on either side of your nose.

Now use your thumb to close your right nostril. Take a slow, deep breath in through your left nostril, counting 6 to 8 seconds. Slow down, you're in-breath so that it takes 6 to 8 seconds to fill your lungs.

Plug your left nostril (both sides blocked now) and hold your breath to a count of 6 to 8.

Now lift your thumb off your right nostril (keep left nostril closed) and breathe out, through your right nostril only, for a count of 6-8.

Do not pause at the end of the breath. Start breathing in through the right nostril to a count of eight.

Plug both sides and hold breath for a count of 6 to 8.

Now breathe out through your left nostril for a count of eight.

Continue to breathe in and out through alternate nostrils, holding your breath on the inhalation only.

IN the left HOLD

OUT the right HOLD

IN the right HOLD

OUT the left HOLD

IN the left HOLD

Breathe with Rhythm that you have measured to personalize this, and the retention of breath is called Kumbhaka.

Retention Breath Control

"The stages are measured with the performance of Puraka (inspiration), Recaka (expiration) and Kumbhaka (retention). The value is calculated by the length of PURAKA, KUMBHAKA, and RECAKA.

That is inhaling, retention, and exhaling. In that order, the assigned values are in seconds.

Pranayama breathing exercises are vital to your life. Pranayama (vital force) awakens Shakti (power), frees it from disease, detaches you from the world, and produces bliss. The respiration, inspiration, and retention will increase over time for each performance. Yoga uses a scale from easy to hard as follows: Adhama – inferior, Madhyama – middle and Uttama – best." -Verse31A1 Tripura-sara-samuccaya, Ch. V, 26

"Kumbhaka is keeping the air confined inside. Rechaka is expelling, exhaling the confined air. The instructions for Puraka (inhale), Kumbhaka, and Rechaka will be found at the proper place, and they should be carefully followed." Gita • Adhama 4, 16, and 8 = 28 Inferior

Inhale for 4 seconds, Hold for 16 seconds, Exhale for 8 seconds = 28

- Madhyama 8, 32, and 16 = 56 Middle
- Uttama 16, 64, and 32 = 112 Best

As you practice, the yogis say that the Adhama causes perspiration and tremors. In Madhyama and Uttama, when done 100 times, it produces levitation.

Hatha Pranayama

Aerobic exercise is an essential need for the body. This Prana exercise is just that. Some of us are not able to do aerobics due to injury or disability. I have used this over many years to open respiration avenues for nasal and increase lung oxygen levels. People who smoke can increase oxygen levels and thoracic size of the chest that have been inhibited by the toxic chemicals of cigarettes. Do this religiously and over time the cells will get more oxygen and nitrogen to help detox inflammatory ingredients, called inactive ingredients.

- First, you must learn to do diaphragm breathing. This means to move the stomach outward from the spine and inhale. Do not continue to draw breath from the upper lung, which greatly reduces the oxygen volume, critical to detoxing the body.
- Second, there are 2 Prana exercises, 1 minute each or less, while holding breath vigorously and stationary sitting or standing, using arms as the central focus of aerobic movement.
- Place your hand on your stomach and notice the breath direction.

Pranayama means breath control taken in a direction. Diaphragm breathing is the correct way to fill your lungs properly. The ancient translation reads that Prana, which dwells in the heart, draws Apana, which dwells in the Muladhara (root chakra), and Apana draws Prana, a falcon attached by a string that is drawn back again when he attempts to fly.

The breathing control is to expand the lower lung by pushing the diaphragm out to fill the lungs from the bottom to the top, and then pushing the diaphragm towards the spine to empty the lungs. When you fill a pitcher with water, the bottom is filled first. The diaphragm is a muscle, and this control will help strengthen the belly. This will increase the oxygen throughout your body dramatically. The air we breathe is the same energy that inspires all physical matter and releases tension. During the Hatha-Yoga postures, breath control is done rhythmically.

- Inhale diaphragmic style and hold your breath. Standing.

Move arms to the middle of the chest with fists clenched and move out away from the body and back into the chest, repeat for 1 minute or less. If you are at 15 seconds, keep practicing until you add 5 seconds to each workout.

Do this vigorously, creating aerobic activity. Moving your arms in and out of your chest. This exercise will be as good as running fast for 1 minute. So, you become a sprinter as far as the respiratory process is now engaged.

Now move your arms, stretching out from your body on both sides and flexing at the elbow. Fists are brought to the shoulder and moved out again to the sides. Holding breath by inhaling and rapidly moving your arms out and in, bending at the elbows. Hold breath and do as long as 1 minute, starting at a lower time and increasing over time!

Move out of this position rapidly!

Move in towards the chest rapidly!

The dynamics of this Breathing Exercise can help you gain the level of a marathon runner as you increase holding breath and rapidly execute the movements for 60 seconds! Repeating 4 times as you practice will increase growth and will help smokers open up lung areas.

Diet and exercise have been well documented for thousands of years, and with modern FDA-approved chemicals and mass-produced food and medicine, a divergent pathway of metabolic brilliance and photon highways of light travel!

Children of Light

From the moment of your creation in the womb of your mother's light from the union of sperm and embryo, there has been a constant level of regeneration with the energy of your parents' DNA, with encoded mathematical illusions discovered by Einstein and further explain nature, where he said you will understand everything better. The master plan is guided by light!

1. The separation between past, present, and future is only an illusion, although a convincing one.
2. This separation between past, present, and future has the value of mere illusion, however tenacious.
3. The distinction between past, present and future is only a stubbornly persistent illusion. "Einstein!"

"Full-spectrum light is light that covers the electromagnetic spectrum from infrared to near-ultraviolet, or all wavelengths that are useful to plant or animal life; in particular, sunlight is considered full-spectrum, even though the solar spectral distribution reaching Earth changes with time of day, latitude, and atmospheric conditions." "Full-spectrum Light Sources." Lighting Answers. Vol. 7, no. 5. March 2005 [September 2003]. Archived from the original on December 12, 2012. Retrieved 2007-11-01.
Nutrition is photosynthesized in plants with 118 Alkaline elements listed on the Periodic Table. Alkaline is light codes sent from the Sun to the plants, making plant medicine far superior to pharmaceuticals simply because it uses photons and not chemicals!

The Universe are particles in an electromagnetic force using particles of elements. "Charged or uncharged subatomic particles. Electromagnetic radiation, on the other hand, consists of massless particles called photons. It is created when a charged atomic particle changes its velocity. Examples of electromagnetic radiation include radio waves, visible light, X-rays, and gamma rays. " https://www.livescience.com

PARTICULATE RADIATION
VERSUS
ELECTROMAGNETIC RADIATION

Visit www.PEDIAA.com

PARTICULATE RADIATION	ELECTROMAGNETIC RADIATION
Particulate radiation involves the emission of particles such as alpha or beta particles, which are charged or uncharged subatomic particles	Electromagnetic radiation consists of photons, which are massless particles with no charge
Particulate radiation particles may carry a positive (alpha) or negative (beta) charge	Electromagnetic radiation, such as light or radio waves, carries no net charge
Involves physical particles	Characterized by wave-like behavior
Tends to have lower penetration ability	Tends to have higher penetration ability

"Electromagnetic radiation - Spectrum, Wavelengths, Properties.

Electromagnetic radiation is produced whenever a charged particle, such as an electron, changes its velocity—i.e., whenever it is accelerated or decelerated. Properties and behavior of electromagnetic radiation.

Examples include radio waves, infrared radiation, visible light, ultraviolet radiation, X-rays, and gamma rays. Electromagnetic radiation exhibits wavelike properties." https://www.britannica.com › science › electromagnetic-radiation › The-electromagnetic spectrum

This universal energy from light that encodes the universe with regeneration from light in an electromagnetic particle is fused into plants and animals for enlightened physics. Again, I want to point out Einstein's discovery of concentrated energy on the atomic level. Of illusional mathematics. Light codes drawn to and through us and all the universe with magnetic and photonic particles of energy.

"The brief account of familiar phenomena given above surveyed electromagnetic radiation from low frequencies of ν (radio waves) to exceedingly high values of ν (gamma rays). Going from the v values of radio waves to those of visible light is like comparing the thickness of this page with the distance of Earth from the Sun, which represents an increase by a factor of a million billion. Similarly, going from the v values of visible light to the very much larger ones of gamma rays represents another increase in frequency by a factor of a million billion. This extremely large range of v values, called the electromagnetic spectrum, is shown in Figure 1, together with the common names used for its various parts, or regions.

That's a vivid and powerful analogy—comparing the leap in frequency from radio waves to visible light to the vast gulf between a page's thickness and the Earth–Sun distance really drives home the staggering scale of the electromagnetic spectrum." https://pediaa.com/what-is-the-difference-between-particulateand-electromagnetic-radiation/

This amazing power of a million billion repeated million billion to an invisible light! This speed is what plant-based energy consumption uses to run your organic battery! "Visible light, the narrow portion of the spectrum perceptible to the human eye, plays a crucial role in our daily lives. Beyond illuminating the world around us, it is essential for photosynthesis in plants, allowing them to convert light energy into chemical energy." https://pediaa.com/what-is-thedifference-between-particulate-and-electromagnetic-radiation/

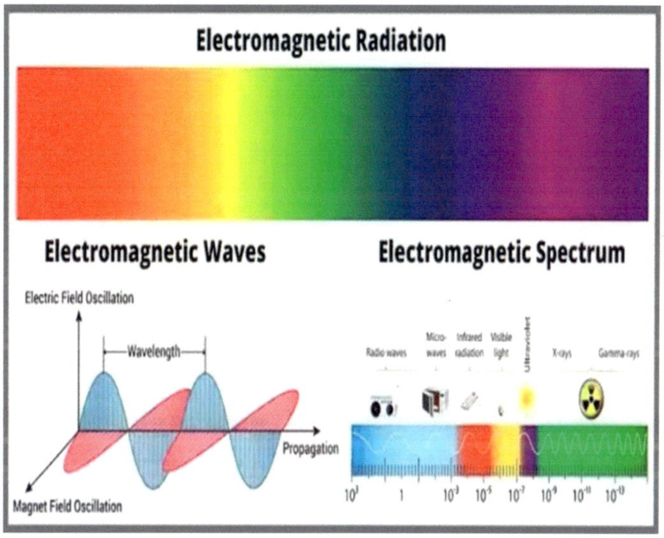

The periodic table shows 118 elements that are fused into plants and animals.

PERIODIC TABLE

This is the common long form of the table. Roman numerals and letters heading the vertical columns indicate the groups. (There are differences of opinion regarding the letter designations, but those given here are probably the most generally used. Also, international standards favor numbering the groups 1–18 from left to right using Arabic numerals, but the designations shown below remain quite common.) The horizontal rows represent the periods, with two series removed from the two very long periods and represented below the main table. Atomic numbers are given above the symbols for the elements. Compare ELEMENT table.

IA[1]	IIA[2]	IIIB	IVB	VB	VIB	VIIB	VIII	VIII	VIII	IB	IIB	IIIA	IVA	VA	VIA	VIIA[3]	VIIIA[4]
1 H																1 H	2 He
3 Li	4 Be											5 B	6 C	7 N	8 O	9 F	10 Ne
11 Na	12 Mg											13 Al	14 Si	15 P	16 S	17 Cl	18 Ar
19 K	20 Ca	21 Sc	22 Ti	23 V	24 Cr	25 Mn	26 Fe	27 Co	28 Ni	29 Cu	30 Zn	31 Ga	32 Ge	33 As	34 Se	35 Br	36 Kr
37 Rb	38 Sr	39 Y	40 Zr	41 Nb	42 Mo	43 Tc	44 Ru	45 Rh	46 Pd	47 Ag	48 Cd	49 In	50 Sn	51 Sb	52 Te	53 I	54 Xe
55 Cs	56 Ba	57 *La	72 Hf	73 Ta	74 W	75 Re	76 Os	77 Ir	78 Pt	79 Au	80 Hg	81 Tl	82 Pb	83 Bi	84 Po	85 At	86 Rn
87 Fr	88 Ra	89 #Ac	104 Rf	105 Db	106 Sg	107 Bh	108 Hs	109 Mt	110 Ds	111 Rg	112 Cn	113 Nh	114 Fl	115 Mc	116 Lv	117 Ts	118 Og

* LANTHANIDE SERIES	58 Ce	59 Pr	60 Nd	61 Pm	62 Sm	63 Eu	64 Gd	65 Tb	66 Dy	67 Ho	68 Er	69 Tm	70 Yb	71 Lu
# ACTINIDE SERIES	90 Th	91 Pa	92 U	93 Np	94 Pu	95 Am	96 Cm	97 Bk	98 Cf	99 Es	100 Fm	101 Md	102 No	103 Lr

[1]Group IA (excluding hydrogen) comprises the alkali metals.
[2]Group IIA comprises the alkaline earth metals.
[3]Group VIIA (excluding hydrogen) comprises the halogens.
[4]Group VIIIA (also called group Zero) comprises the noble gases.

Research has demonstrated that light permeates all matter in the universe, an achievement recognized with the awarding of the Nobel Prize in Medicine. to Einstein and fuses solar elements into plants that are essential for nutrition. The earth and minerals are a product of fire in the center of the earth. As the molten lava cooled in the oxygen, hydrogen and solar radiation, the DNA made rocks and jewels, and a terra-forest where petrified rocks formed! This then formed into protein energy from the aforementioned gases, and life began with bacteria!

The bacteria are spread to the DNA from RNA, and the type of bacteria that makes a difference is antigens or anti-bacteria that are probiotic from alkaline nutrients based on pollination of the plant world. You have taken antibiotics, but the body makes them! Along with illusional mathematics that was founded in 1915 by Einstein proved the atomic level of cells energy is transmitted between cells with an alkaline code, RNA, and downloaded on DNA, fulfilling organ structure in a maze of protein structures for health and intelligence.

Understanding this can be illustrated by shopping! The barcode on the product is read by a scanner, and the product information and price! Your body scans the RNA from plants and then applies the value to the DNA for genetic regeneration!

Pharmaceuticals are chemical syntheses and plants, a Solar Radiated synthesis (Photosynthesis)

Cultivating this with hydrogen, nitrogen from oxygen and Solar Radiated Universe, the light can increase with frequency and spectrum to stimulate electromagnetic activity passing RNA codes of mega levels of intercellular transmission! This regenerates by cell-

neutralizing and medicines are cell-killing, indiscriminate of healthy cells and toxic growth that cause side effects with other cells. They are not biological, or life logical, as illusional mathematics of photons communication of the Body of Light.

Plants are the key to this level that have the root level of RNA, Ribonucleic Acid, with Glycol sugars, best as organic. This data is transmitted to DNA, which consists of deoxyribose nucleic acids. Oxygen breaks down, so the nucleic energy in the cell is preserved! Processed sugar rides the energy cycle, lacks regeneration with false information to energy centers, whereas the organic sugar rebuilds and sustains energy.

DNA, "Deoxyribonucleic acid (/diːˈɒksɪˌraɪboʊnjuːˌkliːɪk, -ˌkleɪ-/ ⓘ;[1] DNA) is a polymer composed of two polynucleotide chains that coil around each other to form a double helix. The polymer carries genetic instructions for the development, functioning, growth and reproduction of all known organisms and many viruses. DNA and ribonucleic acid (RNA) are nucleic acids. Alongside proteins, lipids and complex carbohydrates (polysaccharides), nucleic acids are one of the four major types of macromolecules that are essential for all known forms of life." Alberts B, Johnson A, Lewis J, Raff M, Roberts K, Walter P (2014). Molecular Biology of the Cell (6th ed.). Garland. p. Chapter 4: DNA, Chromosomes and Genomes. ISBN 978-0-8153-4432-2. Archived from the original on 14 July 2014.

Your body is biological, neurological, and electromagnetic. These terms refer to the action of the spectrum energy transfer that is logical to metabolic pathways! With this structure of protein complexes, we have intercellular communication that can translate nature's energy codes with understanding or logical to the cells' energy path! Consumption of organic plants achieves this with millions of billions in 26 trillion cells drawn magnetically with electric connectors.

Eating organic plants boosts immunity and uses electromagnetics to draw pollination in plants that carry nutrients from digested nutrients! This is how a spectrum of energy broadens and then narrows for energy propelling with phosphates to a full spectrum of light codes in RNA photosynthesized nutrients! As the magnetic and electric energy is moving at the speed of light, the RNA is downloaded on the DNA for genetic and phonetic expression of your body structure.

What people should realize is their Body Light, and we see the rest of nature as an illusion or light waves gathered by optical functions and gathered in memory and brain functions for thought and command of homeostasis regulatory functions controlling bodily functions!

Mother Earth and Father Sky are a metaphorical phrase by Native Americans! They knew about this from RNA/ DNA information of dynamic energy, as we do when enlightened to the vast, incredible power within and without us!

Below is a picture of DNA and the elements that gather for energy construction.

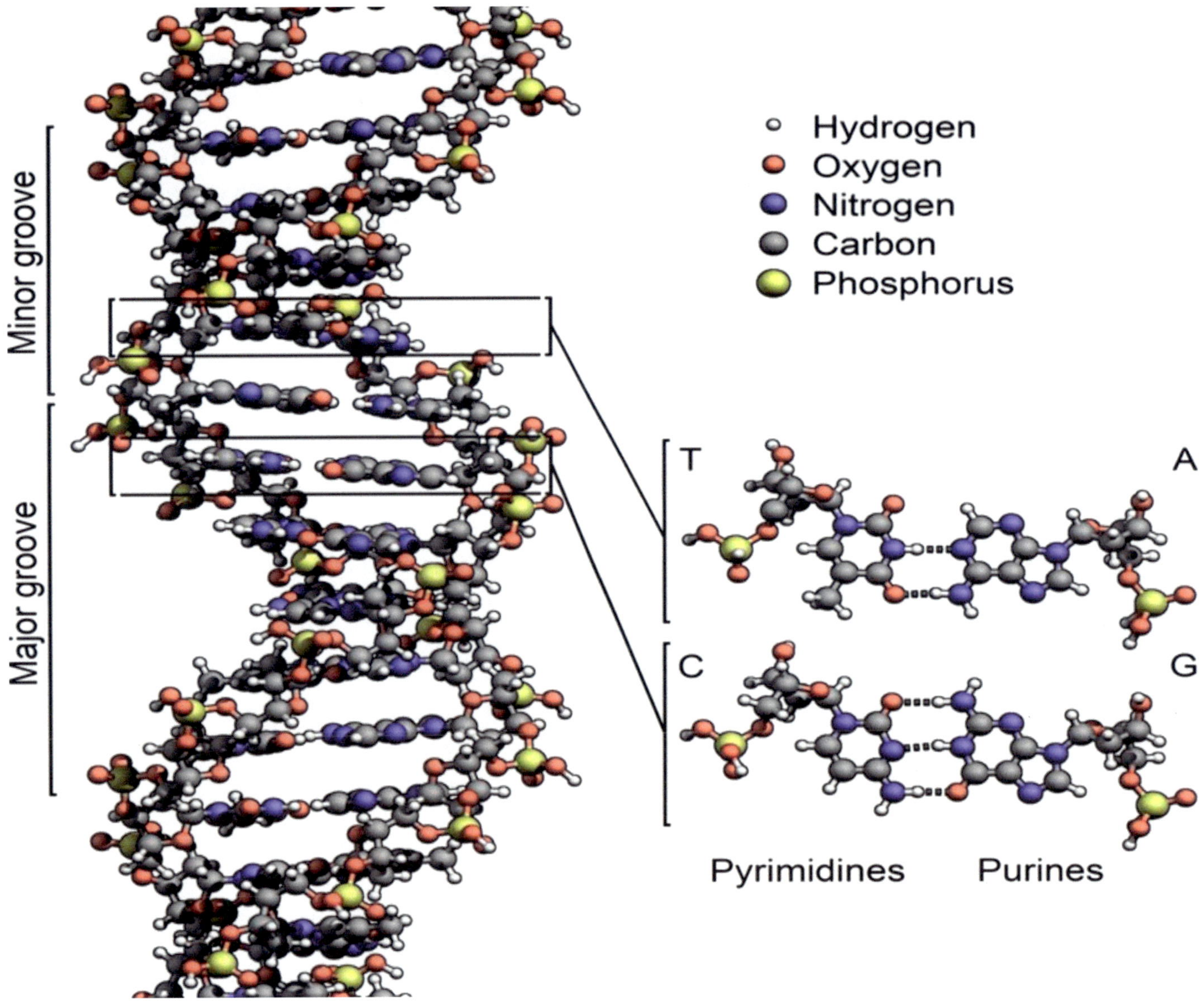

"The creator has a master plan, happiness for every man. Peace and harmony for every man." Pharaoh Sanders Karma Album

Dendrites and synapses facilitate electrical connectivity. "Dendrites receive signals from other neurons, while synapses are the junctions where communication occurs between neurons, playing a crucial role in neural signaling.

Dendrites:

Structure: Dendrites are tree-like extensions that branch out from the cell body (soma) of a neuron. They increase the surface area available for receiving signals from other neurons.

Function: Their primary role is to receive electrochemical signals from the axon terminals of other neurons through synapses. Dendrites convert these chemical signals into electrical impulses that are transmitted toward the soma.

Synaptic Connections: Dendrites are covered with synapses, which are the contact points where information from presynaptic neurons arrive. The extensive branching of dendrites allows a single neuron to receive inputs from thousands of other neurons. "https://en.wikipedia.org/wiki/Dendrite

By Millencolin - Own work, CC BY-SA 4.0, https://commons.wikimedia.org/w/index.php?curid=136366067

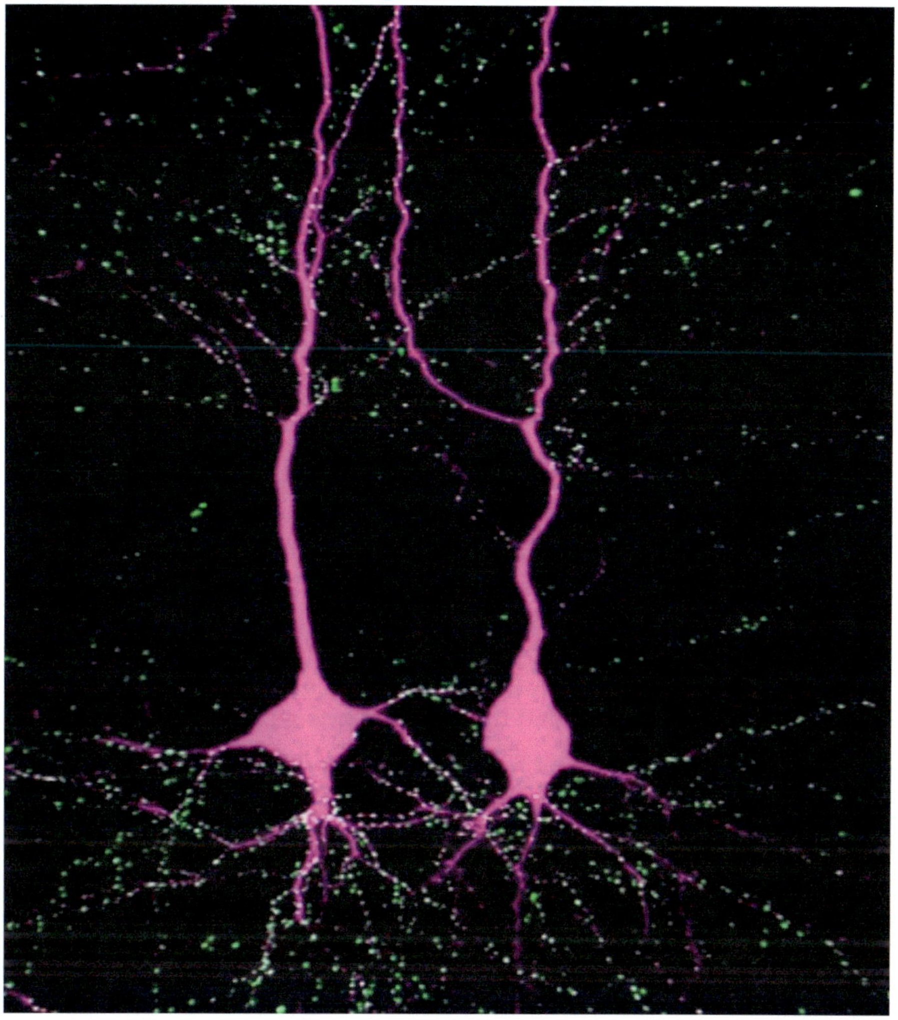

Summation:

Your enlightened body needs the harmony of light, a full spectrum, that is electrically charged!

Children of Light and Spectrum Harmony

The Origins of Light and Regeneration

From the very beginning of life, light has played a constant role in regeneration, encoded through mathematical principles discovered by Einstein. Einstein highlighted the illusionary nature of time, emphasizing that the separation between past, present, and future is merely a persistent illusion.

Nature and Power of Full-Spectrum Light

Full-spectrum light encompasses wavelengths from infrared to near-ultraviolet, including all that are beneficial to plant and animal life. Sunlight is considered full-spectrum, and its distribution varies with time, latitude, and atmospheric conditions.

Photosynthesis and Nutrition

Plants synthesize nutrition through the process of photosynthesis, utilizing 118 alkaline elements from the Periodic Table. These elements, energized by sunlight, make plant-based medicine superior to pharmaceuticals.

Electromagnetic Forces in the Universe

The universe operates through electromagnetic forces and elemental particles. Electromagnetic radiation is produced when charged particles, such as electrons, are accelerated. This radiation includes radio waves, visible light, X-rays, and gamma rays, and consists of massless photons.

Energy Transmission and Scale

The electromagnetic spectrum spans an enormous range of frequencies. The leap from radio waves to visible light is as vast as comparing the thickness of a page to the distance from Earth to the Sun. This immense scale demonstrates the staggering power and reach of electromagnetic energy. A million billion followed by another million billion light transmissions to visible and invisible illusions.

Elements and Life

The periodic table's 118 elements are fused into both plants and animals. Light from the universe permeates all matter, a principle proven by Einstein's Nobel-winning research. Solar elements fuse into plants, which become essential nutrients. Earth's minerals originate from fire at the planet's core, and as molten lava cooled in the presence of oxygen, hydrogen, and solar radiation, DNA formed rocks, minerals, and jewels, leading to the rise of bacteria and life itself.

Role of Bacteria and Genetic Codes

Bacteria spread genetic information via RNA, and certain types are beneficial, such as probiotics from alkaline nutrients found in pollinated plants. The body naturally produces antibiotics, and the transmission of cellular energy operates through an alkaline code, RNA, and DNA, supporting organ structure and health.

Illustrative Analogy: Scanning and Regeneration

Just as a barcode scanner reads product information, the body scans RNA from plants and applies its value to DNA for genetic regeneration. Pharmaceuticals are chemically synthesized, while plants undergo solar radiated synthesis (photosynthesis).

Energy Cultivation

With the interplay of hydrogen, nitrogen, oxygen, and solar radiation, light frequency stimulates electromagnetic activity, enabling RNA codes to transmit high levels of intercellular information. Organic plant consumption regenerates cells, while synthetic medicines indiscriminately kill cells, causing side effects. Plant-Based Nutrients and Genetic Expression

Plants contain RNA (ribonucleic acid with glycol sugars), which is best received from organic sources. This information is passed on to DNA (deoxyribonucleic acid), preserving cellular energy. While processed sugars provide energy without regeneration, organic sugars rebuild and sustain energy.

Biological and Electromagnetic Foundations

Human bodies are biological, neurological, and electromagnetic, facilitating logical energy transfer through metabolic pathways. Proteins enable intercellular communication, translating nature's energy to support health. Organic plants help build immunity and support energy cultivation.

Spectrum of Energy and Genetic Download

Organic plant nutrients expand the body's energy spectrum, using phosphates and light codes in RNA for photosynthesis. As magnetic and electric energy travels at light speed, RNA is downloaded onto DNA for genetic and phonetic expression.

Light, Illusion, and Homeostasis

The body's light is an essential force, while the rest of nature appears as an illusion, gathered by optical and neurological functions that support memory and bodily regulation.

Metaphorical Understanding

Native Americans referred to "Mother Earth and Father Sky" to encapsulate the dynamic energy exchanged between RNA and DNA. Recognizing this vast power illuminates our connection to the universe.

The Need for Harmony

As referenced in Pharoah Sanders' Karma Album, happiness, peace, and harmony are part of the creator's master plan. Achieving harmony requires the enlightened body to embrace the full spectrum of electrically charged light.

The Origins of Light and Regeneration

From the moment of your creation—within your mother's womb, formed through the union of sperm and embryo—regeneration has been a constant process. This renewal is driven by the energy encoded in your parents' DNA, following mathematical principles discovered by Einstein. Einstein's work revealed the illusory nature of time and emphasized that our understanding of the past, present, and future is only a convincing illusion. The master plan of existence is guided by light, as Einstein noted:

- The separation between past, present, and future is only an illusion, although a convincing one.
- This separation has the value of mere illusion, however tenacious.
- The distinction between past, present, and future is only a stubbornly persistent illusion.

Nature and Power of Full-Spectrum Light

Full-spectrum light is defined as light covering the electromagnetic spectrum from infrared to near-ultraviolet—all wavelengths useful to plant and animal life. Sunlight is considered full spectrum, although its spectral distribution reaching Earth changes with time of day, latitude, and atmospheric conditions.

Photosynthesis and Nutrition

Nutrition in plants is produced by photosynthesis, which utilizes 118 alkaline elements found on the Periodic Table. These elements, energized by sunlight, make plant-based medicine far superior to pharmaceuticals. Alkaline light codes from the Sun are transmitted to plants, making them powerful sources of healing.

Electromagnetic Forces in the Universe

The universe is composed of particles governed by electromagnetic forces.
Electromagnetic radiation is generated when charged particles, such as electrons, are accelerated or decelerated. This radiation consists of massless photons and includes radio waves, visible light, X-rays, and gamma rays. The movement of charged particles produces oscillating electromagnetic fields, which are fundamental to the universe's structure and behavior.

Energy Transmission and Scale

The electromagnetic spectrum encompasses an incredibly broad range of frequencies. Moving from the frequencies of radio waves to visible light is comparable to the difference between the thickness of a page and the distance from Earth to the Sun, a leap by a factor of a million billion. The jump from visible light to gamma rays is equally vast. This analogy illustrates the immense scale and power of electromagnetic energy.

This power, repeated by factors of a million billion, is harnessed by plants through energy consumption, fueling organic life as a kind of natural "battery."

Elements and Life

The periodic table's 118 elements are fundamental to both plants and animals. Research has shown that light permeates all matter, a discovery recognized by the awarding of the Nobel Prize in Medicine to Einstein. Solar elements fuse into plants, providing essential nutrition. The minerals of the Earth are products of the planet's fiery core. As molten lava cooled amid oxygen, hydrogen, and solar radiation, DNA formed rocks, jewels, and forests—eventually giving rise to bacteria and, ultimately, life.

Role of Bacteria and Genetic Codes

Bacteria distribute genetic information through RNA, and certain types—such as probiotics from alkaline nutrients in pollinated plants—are beneficial. The body produces its own antibiotics, and the transmission of energy between cells occurs via an alkaline code, RNA, and DNA. This process is essential for the formation and health of organ structures through a network of proteins.

Illustrative Analogy: Scanning and Regeneration

Understanding this process can be likened to shopping: a barcode on a product is read by a scanner, revealing product information and price. Similarly, the body scans RNA from plants and applies this information to DNA, enabling genetic regeneration.

Pharmaceuticals vs. Plant-Based Synthesis

Pharmaceuticals are the result of chemical synthesis, whereas plants undergo solar radiated synthesis, or photosynthesis. Plants combine hydrogen and nitrogen from oxygen with solar energy, increasing light frequency and stimulating electromagnetic activity. This process allows RNA codes to facilitate high-level intercellular communication and regeneration. In contrast, synthetic medicines indiscriminately kill both healthy and unhealthy cells, often causing side effects because they are not biological or "life logical" in the way plant-derived nutrients are.

Plant-Based Nutrients and Genetic Expression

Plants contain RNA (ribonucleic acid with glycol sugars), which is most beneficial when sourced organically. This information is transferred to DNA (deoxyribonucleic acid), preserving cellular energy. While processed sugars provide energy without true regeneration, organic sugars support and sustain energy by rebuilding it.

DNA, a polymer of two polynucleotide chains forming a double helix, carries genetic instructions for growth, development, and reproduction in all known organisms. Alongside proteins, lipids, and complex carbohydrates, nucleic acids are essential macromolecules for life.

Biological and Electromagnetic Foundations

The human body functions as a biological, neurological, and electromagnetic system. This means that energy transfer, guided by the spectrum of light, is logical to metabolic pathways. Proteins facilitate intercellular communication, translating nature's energy to ensure health and intelligence. Consuming organic plants builds immunity through energy cultivation.

Eating organic plants broadens the body's energy spectrum. Nutrients are carried by pollinated plants and digested before being propelled through energy pathways using phosphates and full-spectrum light codes in RNA. As magnetic and electric energy moves at the speed of light, RNA is downloaded onto DNA, enabling genetic and phonetic expression of the body's structure.

Light, Illusion, and Homeostasis

The human body is fundamentally a "body of light," while the rest of nature is perceived as an illusion of light waves, gathered by optical and neurological processes. These processes support memory, thought, and the homeostatic regulation of bodily functions.

Metaphorical Understanding

Native Americans used the phrase “Mother Earth and Father Sky” as a metaphor for the dynamic energy exchange between RNA and DNA. Awareness of this energy reveals the immense power both within and around us.

The Need for Harmony

As Pharaoh Sanders’ Karma Album expresses, “The creator has a master plan, happiness for every man. Peace and harmony for every man.” Achieving harmony requires the enlightened body to embrace a full spectrum of electrically charged light, leading to regeneration, health, and fulfillment.

To illustrate the truth of this, you need to be in a dark room at night with one another and turn off all the lights. You will not be able to see each other until the lights are restored, and light is then transferred to your optical function. An illusion!

Addendum:

For in-depth routines and 100 positions, meditation, and Chakra information, you can order online Yoga Light! Enter Glenn Moran Yoga Light Book

Yoga Light is a scientific, spiritual and physiological journey of ancient wisdom that will give you energy and strength through the natural process of yoga. There are full instructions and inspiring reflections, as well as 8 routines to practice that bring bliss.

For online classes to help with an exercise program, here are Andrea Tejas' online classes.

www.infin8space.com

Yoga Light: Nature, Energy & Strength: Moran, Glenn: 9781495452741: Amazon.com: Books

Made in the USA
Las Vegas, NV
03 November 2026

57424298R00052